Hepatitis C Virus Treatment

Kazuaki Chayama
Editor

Hepatitis C Virus Treatment

Highly Effective Therapy with Direct Acting Antivirals and Associated Viral Resistance

Springer

Editor
Kazuaki Chayama
Department of Gastroenterology
and Metabolism
Hiroshima University
Hiroshima, Japan

ISBN 978-981-10-2415-3 ISBN 978-981-10-2416-0 (eBook)
DOI 10.1007/978-981-10-2416-0

Library of Congress Control Number: 2016956568

Printed on acid-free paper

This Springer imprint is published by Springer Nature
The registered company is Springer Nature Singapore Pte Ltd.
The registered company address is: 152 Beach Road, #22-06/08 Gateway East, Singapore 189721, Singapore

Preface

Interferon has served for many years as the backbone of chronic hepatitis C therapy, despite the dismal success rate of early therapies (<10% for genotype 1b). While the introduction of peg-interferon and the addition of ribavirin improved the sustained viral response (SVR) rate, the therapy was still ineffective for about half of the patients infected with genotype 1b, providing a strong impetus for the development of innovative therapeutic options for HCV. In stark contrast to the system-wide innate immune activation induced by interferon, the next major advance in HCV therapy employed a far more directed strategy. Direct acting antiviral agents (DAAs) have been designed to interfere with HCV replication by directly targeting HCV proteins. Administered as part of a triple therapy along with peg-interferon plus ribavirin, protease inhibitors, the first DAAs to be approved, significantly improved the SVR rate. However, interferon's broad activity has always exacted a high cost in terms of side effects, including fever, malaise, anorexia, and thrombocytopenia, and ribavirin poses a high risk of anemia, as well, requiring frequent dose reductions. These risks put interferon-based therapy out of the reach of many of Japan's patients, many of whom are elderly. Furthermore, patients with mental illness, cirrhosis, or other comorbidities are ineligible for interferon-containing therapies. However, protease inhibitors used in monotherapy would rapidly select for antiviral resistance and lead to viral breakthrough and discontinuation of therapy. The race was on for a safe and effective alternative to interferon. Studies using the human hepatocyte chimeric mouse as a model system for HCV infection demonstrated that the combination of two DAAs with different targets (telaprevir, an NS3/4A protease inhibitor, and MK-0608, an NS5B polymerase inhibitor), could successfully eliminate HCV without interferon or ribavirin. Clinical trials in humans using the combination of a protease inhibitor (asunaprevir) and an NS5A inhibitor (daclatasvir) showed great success, especially against genotype 1b, the predominant genotype in Japan. In 2014, Japan became the first country to approve an interferon-free therapy for HCV and has since successfully treated tens of thousands of patients. Not all patients achieve SVR, however, and pre-existing or treatment-emergent resistance-associated variants contribute to treatment failure in an important subset of patients. Therefore, clinicians attempting to apply DAA therapy treatment must understand the effects of each drug and be cognizant of their resistance characteristics. This special issue outlines current and future treatments for HCV with the goal of familiarizing medical personnel with issues in DAA treatment for hepatitis C.

Hiroshima, Japan Kazuaki Chayama

Contents

1 Treatment of Chronic Hepatitis C with the First-Generation Protease Inhibitor Telaprevir: Its Efficacy and Resistance Mutations

Yoshiyasu Karino

Abstract

Telaprevir, a first-generation protease inhibitor, was approved in 2011 for use in antiviral therapy for hepatitis C in combination with PEG-IFN and ribavirin, and treatment of hepatitis C entered a new stage. In the Japanese phase III trial, triple therapy with telaprevir/PEG-IFN/ribavirin showed a much higher sustained viral response (SVR) rate (73 %) than PEG-IFN and ribavirin combination alone (49.2 %) in treatment-naïve patients. Furthermore, in clinical practice more than 90 % of treatment-naïve patients achieved SVR by management of drug dosing. In most cases, telaprevir-resistant variants appeared at the time of the treatment failure. But most telaprevir-resistant strains were replaced by wild-type HCV in the natural course. Now, in the era of IFN-free therapy, the role of TVR has decreased, but TVR played a key pioneering role in the shift to direct-acting antiviral (DAA) therapy.

Keywords

Telaprevir • Triple therapy • Resistance mutation

1.1 Introduction

Antiviral treatment for hepatitis C started with interferon (IFN) monotherapy in 1992. In 2001, combination therapy with ribavirin (RBV) became available. In 2004, combination therapy with pegylated interferon (PEG-IFN) alpha and RBV for 48 weeks became the standard therapy for patients infected with hepatitis C virus

Y. Karino (✉)
Department of Hepatology, Sapporo-Kosei Hospital,
North-3 East-8, Chuou-ku, Sapporo 060-0033, Japan
e-mail: ykarino@ja-hokkaidoukouseiren.or.jp

K. Chayama (ed.), *Hepatitis C Virus Treatment*,
DOI 10.1007/978-981-10-2416-0_1

(HCV) genotype 1 with high viral load, considered among the most difficult to treat. However, the rate of sustained viral response (SVR) was reported to be 50 % or less after combination therapy with PEG-IFN/RBV and 60 % or less after extended therapy for 72 weeks. Recently, new drugs known as direct-acting antivirals (DAAs), which target viral proteins directly, have been developed. The major DAAs that are under development are NS3/NS4A protease inhibitors, NS5A inhibitors, and NS5B polymerase inhibitors. The protease inhibitors were the first to enter clinical trials. In May 2011, telaprevir (TVR) [1–4] by Vertex and boceprevir [5] by Merck were approved by the Food and Drug Administration (FDA) in the United States. TVR was approved in Japan in September 2011, and antiviral treatment of HCV entered a new stage. Protease inhibitors can be divided into acyclic (linear) and macrocyclic types based on structure. TVR, belonging to the former type, is a first-generation protease inhibitor, and simeprevir [6] and subsequently developed drugs belonging to the latter group are called second-generation inhibitors. The present study focuses on the efficacy of TVR, a first-generation protease inhibitor, in patients infected with HCV.

1.2 Protease Inhibitors

HCV has a single-stranded RNA genome of 9.6 kb in length that encodes structural and nonstructural (NS) proteins. The HCV genome encodes a polyprotein of approximately 3000 amino acids that is processed by cellular and viral proteases to 10 polypeptides. The NS3 has serine protease activity, which is essential for virus replication. The NS3-5 polyprotein is processed by the viral serine protease into nonstructural proteins. NS3/NS4 protease inhibitors inhibit enzymes by binding to the active sites, leading to the direct inhibition of HCV replication.

1.3 Telaprevir

1.3.1 Clinical Trial Results

In 2004, the first clinical trial of TVR as monotherapy was conducted in patients infected with high viral load of genotype 1 HCV. Although HCV RNA levels were significantly decreased in the early stage of treatment, a high proportion of patients experienced viral breakthrough (VBT), indicating treatment failure due to selection for resistance-associated variants. Analysis of serum samples after VBT identified four mutations in the NS3 region (181 amino acids) showing different levels of resistance to TVR [3]. As a result, the approach was shifted to triple therapy of TVR in combination with PEG-IFN/RBV.

In Japan, a phase I TVR clinical trial was started in December 2007. At first, a trial of TVR monotherapy was conducted, but the majority of subjects showed resistance mutations and did not achieve SVR. As in the overseas studies, the approach was shifted to TVR/PEG-IFN/RBV triple therapy. A dose-comparison study was

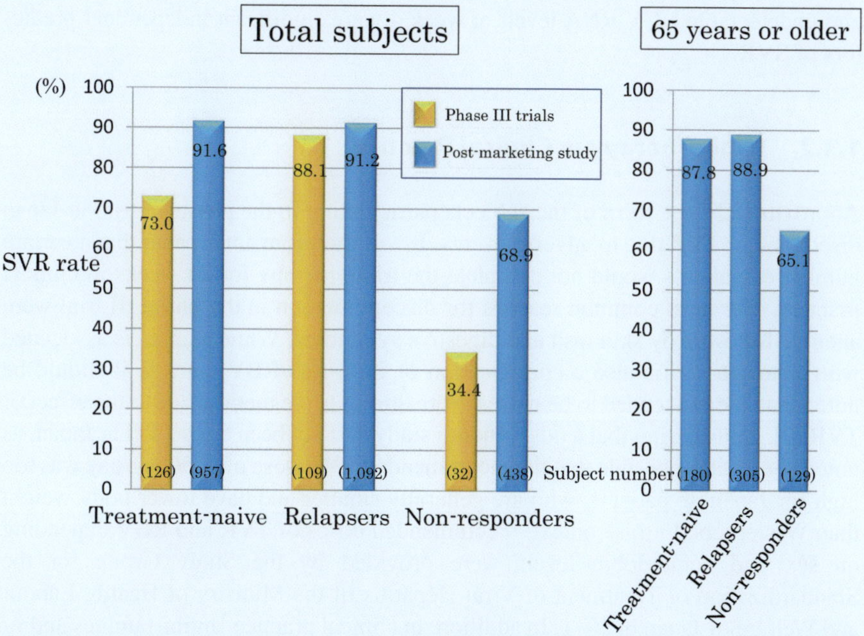

Fig. 1.1 SVR rates of triple therapy with telaprevir (the results from clinical trials and a post-marketing study)

not undertaken in order to avoid drug lag. In the phase III trials, patients, who included interferon treatment-naïve patients, relapsers to prior therapy, and nonresponders to prior therapy, were treated with TVR triple therapy. A total of 2250 mg/day of TVR was administered at 750 mg doses three times per day at an interval of 8 h as in the overseas trials. Patients infected with high viral load chronic HCV genotype 1 received a 12-week course of the triple therapy, followed by a 12-week course of PEG-IFNα-2b/RBV combination therapy (T12/PR24). A comparison study between T12/PR24 and a 48-week course of treatment with PEG-IFNα-2b/RBV (PR) was performed only for the treatment-naïve patients. The SVR rates were 49.2 % (PR48) and 73.0 % (T12PR24) in treatment-naïve patients, 88.1 % in relapsers (T12/PR24), and 34.4 % in nonresponders (T12/PR24) (Fig. 1.1). The analysis showed that the rate of SVR following triple therapy was high in treatment-naïve patients and relapsers but was less than 40 % in the prior nonresponders. Regarding the safety evaluation of 267 patients who began triple therapy, 91 patients (34.1 %) discontinued TVR due to adverse events, including anemia or decreased hemoglobin (15.4 %) and skin rash symptoms (7.1 %) [7]. Chayama et al. examined background and treatment-related factors affecting SVR using multivariate analysis in 94 HCV patients participating in a phase III trial of the triple therapy conducted at the Sapporo-Kosei Hospital, Toranomon Hospital, and Hiroshima University Hospital [8]. They reported that prior treatment response, IL28B major genotype, and undetectable HCV RNA levels at week 4 were significant independent predictors of SVR.

1.3.2 Triple Therapy in Clinical Practice

Approximately one-third of the subjects participating in the phase III trial in Japan discontinued TVR due to adverse events. It was therefore anticipated that a certain number of patients would not complete the triple therapy for 24 weeks in clinical practice. The most common reasons for discontinuation in the phase III trial were anemia, followed by skin rash and digestive symptoms. While anemia is associated with ribavirin and is also a complication of PEG-IFN/RBV therapy, it should be noted that anemia tended to be more severe during triple therapy due to the effect of TVR [9]. Considering that a dose-finding study had not been conducted in Japan, as noted above, it is possible that the recommended TVR dose of 2250 mg/day was too high for Japanese patients, who are generally smaller and have lower body weight than Westerners. In this context, recommended doses of TVR and RBV depending on sex and hemoglobin levels were provided by the Study Group for the Standardization of Treatment of Viral Hepatitis of the Ministry of Health, Labour and Welfare of Japan in 2011. In addition, in clinical practice, antihistamines and/or oral steroids were immediately given when patients developed skin rash. We consider that the above mentioned dose adjustment of TVR and RBV and successful management of adverse drug reactions contributed to the improvement in the SVR rate. A midterm post-marketing study (Vol. 6) was performed in 3563 patients receiving 250 mg Telavic™ (TVR) tablets (efficacy analysis set, 2487). There were more women (59.5 %) than men, and the mean age was 57.7 years old. The most common age group was 55–65 years (43.8 %), and those aged 65 years or older (this age group was not included in the clinical trials) accounted for 24.6 %. The age distribution of the subjects in our hospital is shown in Fig. 1.2. The mean age was 53 years in the IFN monotherapy group, 55 years in the IFN+RBV combination therapy group, and 60 years in the PEG-IFN+RBV+protease inhibitor (PI) group, the last of which was apparently older than the other two groups. The analysis showed that the rate of SVR was 87.5 % overall, 91.6 % in treatment-naïve patients, 91.2 % in relapsers, and 68.9 % in nonresponders, the results of which were far better than those in the clinical trials. In addition, the patients 65 years or older, who had been excluded from the clinical trials, also showed favorable results with SVR rates of 87.8 % in treatment-naïve patients, 88.9 % in relapsers, and 65.1 % in nonresponders (Fig. 1.1).

However, as in clinical trials, incidence of adverse events was high in clinical practice at 96.17 %, and that of severe adverse events including skin disorders, anemia, and renal dysfunction was high at 34.98 %, suggesting that the treatment itself was very hard for patients. Nonetheless, good treatment results were obtained in clinical practice. We consider that the successful clinical outcome was partly attributable to efforts and commitment of our healthcare team and patients.

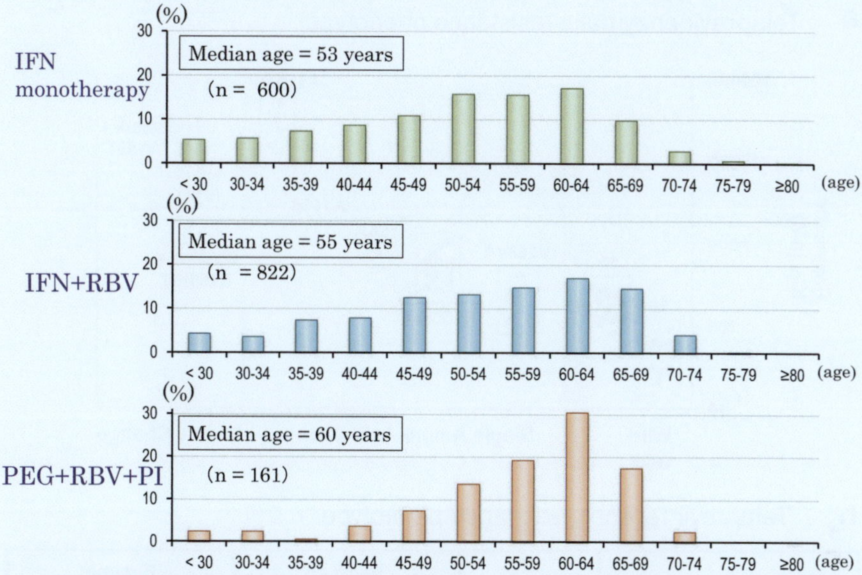

Fig. 1.2 The age distribution of patients receiving IFN monotherapy, IFN+RBV, and IFN+RBV+PI (Department of Hepatology, Sapporo-Kosei Hospital)

1.3.3 Triple Therapy for Genotype 2 Patients

In general, the SVR rate of patients with chronic HCV genotype 2 treated with the standard therapy of PEG-IFNα-2b/RBV for 24 weeks is approximately 80 %; however, there is no established treatment for the remaining 20 % of patients who do not achieve SVR with standard therapy [10]. In this context, a clinical trial of triple therapy was conducted in HCV genotype 2 patients who had received IFN or IFN/RBV combination therapy but had failed or relapsed. They received triple therapy for 12 weeks, followed by PEG-IFNα-2b/RBV for 12 weeks (T12PR24) at the same dose and schedule as in the trial for genotype 1. The SVR24 rate was 88 % (95/108) in relapsers and 50 % (5/10) in nonresponders. The adverse event profile was similar to that for patients with genotype 1. The analysis showed that continuation of treatment for up to 24 weeks was an important predictor of SVR in nonresponders: no patients achieved SVR without completing the therapy [11]. Consequently, the triple therapy was approved for patients infected with HCV genotype 2.

1.3.4 Telaprevir Treatment and Resistance Mutations

The emergence of drug resistance has been a problem in the treatment of HCV with DAAs, as in the treatment of hepatitis B virus with nucleotide analogues. Although TVR monotherapy was observed to have a strong anti-HCV effect in the early stage of the treatment, many patients began to show viral breakthrough (VBT) with the

a Telaprevir enzymatic resistance phenotype

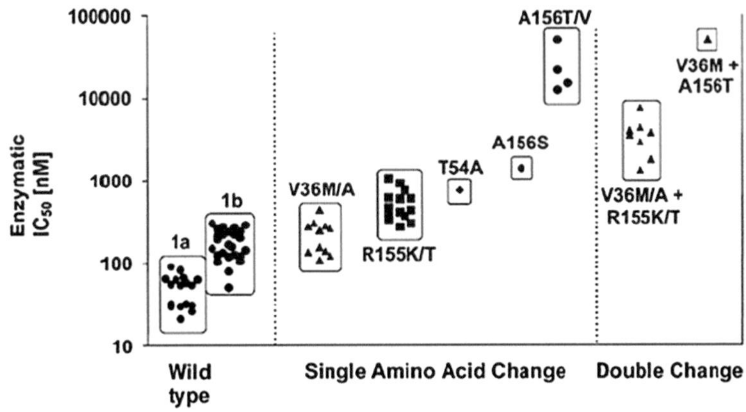

b Telaprevir replicon resistance phenotype

VX-950 Resistance	Mutation	Replicon		Enzyme	
		IC_{50} (fold change)	*N*	IC_{50} (fold change)	*N*
Wild-type	–	1	3	1	1
LOW (<25-fold) Increase in IC_{50}	T54A	6	3	12	1
	V36M	7	3	3	4
	V36A	ND*	3	4.0	5
	R155T	20	3	9.0	6
	R155K	ND	3	8.0	5
	A156S	12	3	22	1
HIGH (>60-fold) Increase in IC_{50}	V36M + R155K	>62	3	71	5
	A156V	>74[31]	3	195, >781	2
	A156T	>74[31]	3	285	2
	V36M + A156T	ND	3	>781	1

*ND= Not determined

Fig. 1.3 Phenotypes of telaprevir-resistant variants. (**a**) Enzymatic IC50 values for telaprevir inhibition of WT and variant NS3 proteases were determined by expressing and purifying the catalytic domain from individual patient clones. The IC50 values are shown on the y-axis. Variants that showed resistance higher than the limit of detection (50 μmol/L) were represented with an IC50 of 50 μmol/L. (**b**) Fold change in replicon and enzymatic IC50 values compared with the HCV reference subtype 1a strain HCV-H are shown for telaprevir inhibition of WT and variant NS3 proteases (This figure is adapted from Fig. 1.3 in Sarrazin et al. [3])

emergence of resistant variants as treatment progressed, resulting in low rates of SVR. Sarrazin et al. reported that mutations that confer low-level resistance (V36A/V36M, T54A, R155K/R155T, and A156S) and high-level resistance (S156V/S156T, V36M/V36A+R155K/R155T, and V36M/V36A+A156T) were detected in patients receiving TVR monotherapy [3] (Fig. 1.3). In addition, Akuta et al.

examined the serum samples of patients who did not achieve SVR with TVR/PEG-IFN/RBV therapy using ultra-deep sequencing. They reported that resistant variants that had been detected in 35 % of pretreatment serum samples were detected only in a small number of the samples from patients experiencing relapse; instead, de novo resistant variants were detected in most patients, including those who did not have resistant variants at baseline [12, 13]. It is also reported that most resistant variants developed from TVR/PEG-IFN/RBV therapy were replaced with wild type within 1 year [14]. In addition, the effectiveness of combination therapy, including the use of a DAA with a different resistance profile, has been reported [15]. We therefore consider that the issue of telaprevir-resistant mutations is manageable.

1.3.5 The Current Role of Triple Therapy with TVR

In 2013, triple therapy including simeprevir (SMV), a second-generation protease inhibitor, was approved. SMV has a significantly lower rate of adverse events compared with TVR and became the first-line treatment for difficult-to-treat patients infected with high viral load HCV genotype 1. In 2014, combination therapy with daclatasvir (DCV) and asunaprevir (ASV) (IFN-free) was approved [16]. Since then, IFN-free therapy has been the mainstay of treatment for HCV genotype 1. In 2015, combination therapy with sofosbuvir (SOF) and RBV was also approved for HCV genotype 2, and it is increasingly used in clinical practice due to its high effectiveness and low rate of adverse events. Given this situation, triple therapy with TVR is indicated for HCV genotype 1 patients who experienced VBT or relapse with DCV/ASV therapy and HCV genotype 2 patients who experienced relapse with SOF/RBV therapy only when IFN is tolerated.

1.4 Conclusion

As we are entering the era of IFN-free therapy for hepatitis C, triple therapy, TVR is used less frequently than in the past. However, TVR played an important role in the development of DAA therapy for HCV.

References

1. Lin K, Perni RB, Kwong AD, Lin C. VX-950, a novel hepatitis C virus (HCV) NS3-4A protease inhibitor, exhibits potent antiviral activities in HCV replicon cells. Antimicrob Agents Chemothr. 2006;50:1813–22.
2. Reesink HW, Zeuzem S, Weegink CJ, Forestier N, van Vliet A, et al. Rapid decline of viral RNA in hepatitis C patients treated with VX-950: a phase Ib, placebo-controlled, randomized study. Gastroenterology. 2006;131:997–1002.
3. Sarrazin C, Kieffer TL, Bartels D, Hanzelka B, Müh U, Welker M, et al. Dynamic hepatitis C virus genotypic and phenotypic changes in patients treated with the protease inhibitor telaprevir. Gastroenterology. 2007;132:1767–77.

4. McHutchison JG, Everson GT, Gordon SC, Jacobson IM, Sulkowski M, Kauffman R, et al. Telaprevir with peginterferon and ribavirin for chronic HCV genotype 1 infection. N Engl J Med. 2009;360:1827–38.
5. Sarrazin, Rouzier R, Wagner F, Forestier N, Larrey D, Gupta SK, et al. SCH503034, a novel hepatitis C virus protease inhibitor, plus pegylated interferon alpha-2b for genotype 1 nonresponders. Gastroenterology. 2007;132:1270–8.
6. Reeink HW, Fanning GC, Farha KA, Weegink C, Van Vliet A, Van't Klooster G, et al. Rapid HCV-RNA decline with once daily TMC435: a phase I study in healthy volunteers and hepatitis C patients. Gastroentelogy. 2010;138:913–21.
7. Kumada H, Toyota J, Okanoue T, Chayama K, Tsubouchi H, Hayashi N, et al. Telaprevir with peginterferon and ribavirin for treatment-naive patients chronically infected with HCV of genotype 1 in Japan. J Hepatol. 2012;56:78–84.
8. Chayama K, Hayes CN, Abe H, Miki D, Ochi H, Karino Y, et al. IL28B but not ITPA polymorphism is predictive of response to pegylated interferon, ribavirin, and telaprevir triple therapy in patients with genotype 1 hepatitis C. J Infect Dis. 2011;204:84–93.
9. Karino Y, Ozeki I, Hige S, Kimura M, Arakawa T, Nakajima T, et al. Telaprevir impairs renal function and increases blood ribavirin concentration during telaprevir/pegylated interferon/ribavirin therapy for chronic hepatitis C. J Viral Hepat. 2013;21(5):341–7.
10. Foster GR, Hézode C, Bronowicki JP, Carosi G, Weiland O, Verlinden L, et al. Telaprevir alone or with peginterferon and ribavirin reduces HCV RNA in patients with chronic genotype 2 but not genotype 3 infections. Gastroenterology. 2011;141:881–9.
11. Kumada H, Sato K, Takehara T, Nakamuta M, Ishigami M, Chayama K, et al. Efficacy of telaprevir-based therapy for difficult-to-treat patients with genotype 2 chronic hepatitis C in Japan. Hepatol Res. 2015;45:745–54.
12. Akuta N, Suzuki F, Seko Y, Kawamura Y, Sezaki H, Suzuki Y, et al. Emergence of telaprevir-resistant variants detected by ultra-deep sequencing in patients infected with HCV genotype 1. J Med Virol. 2013;85:1028–36.
13. Akuta N, Suzuki F, Fukushima T, Kawamura Y, Sezaki H, Suzuki Y, et al. Prediction of treatment efficacy and telaprevir-resistant variants after triple therapy in patients infected with HCV genotype 1. J Clin Microbiol. 2013;51:2862–8.
14. Sullivan JC, De Meyer S, Bartels DJ, Dierynck I, Zhang EZ, Spanks J, et al. Evolution of treatment-emergent resistant variants in telaprevir phase 3 clinical trials. Clin Infect Did. 2013;57:221–9.
15. Sulkowski MS, Gardiner DF, Rodriquez Torres M, Reddy KR, Hassanein T, Jacobson I, et al. Sustained virologic response with daclatasvir plus sofosbuvir +/– ribavirin in chronic HCV genotype 1 infected patients who previously failed telaprevir or boceprevir. J Hepatol. 2013;58:S570.
16. Kumada H, Suzuki Y, Ikeda K, Toyota J, Karino Y, Chayama K, et al. Daclatasvir plus asunaprevir for chronic HCV genotype 1b infection. Hepatology. 2014;59:2083–91.

Features of the Second Wave of the First Generation Protease Inhibitors: Effect and Tolerance

2

Tetsuo Takehara

Abstract

Simeprevir, asunaprevir, vaniprevir, paritaprevir, and other direct-acting antivirals (DAAs) developed after the first-generation protease inhibitors telaprevir and boceprevir are collectively referred to as "second-generation" protease inhibitors. Simeprevir and vaniprevir are used in interferon therapy in combination with pegylated interferon and ribavirin, whereas asunaprevir and paritaprevir are used in interferon-free therapies in combination with the NS5A inhibitors daclatasvir and ombitasvir, respectively. Second-generation drugs are less toxic than first-generation drugs and have different clinical characteristics, including a lower pill burden. Like first-generation drugs, they have a low genetic barrier to resistance, meaning that resistant strains are likely to emerge in nonresponders. However, second-generation drugs have a different resistance profile, with most resistance mutations localized at the D168 residue of the NS3/NS4A protease domain, especially in genotype 1b. It is now known that treatment with asunaprevir after failed treatment with simeprevir tends to be ineffective, illustrating that it is difficult to re-treat patients with a direct-acting antiviral that has the same resistance profile.

Keywords

Simeprevir • Asunaprevir • Vaniprevir • Paritaprevir

T. Takehara (✉)
Department of Gastroenterology and Hepatology, Graduate School of Medicine, Osaka University, Suita, Japan
e-mail: takehara@gh.med.osaka-u.ac.jp

K. Chayama (ed.), *Hepatitis C Virus Treatment*,
DOI 10.1007/978-981-10-2416-0_2

Side Notes

Simeprevir, asunaprevir, vaniprevir, and paritaprevir, which are discussed in this text, as well as faldaprevir, a drug whose development has been discontinued, are all macrocyclic noncovalent inhibitors targeting the NS3/NS4A protease. They have markedly different characteristics than telaprevir and boceprevir, including reduced clinical toxicity and improved pharmacokinetic properties. Nevertheless, they are sometimes still called "second-wave, first-generation" drugs rather than second-generation drugs. This is because these drugs are only effective for treating HCV genotype 1 and have a low genetic barrier to resistance, similar to telaprevir and boceprevir. One example of a true "second-generation" protease inhibitor that has overcome these problems is grazoprevir (MK-5172). Grazoprevir is also a macrocyclic noncovalent inhibitor with good pharmacokinetic properties and is associated with a low rate of resistance even in monotherapy and is effective for a wide variety of genotypes. The A156 residue of the NS3/NS4A protease domain is a known site of mutations that confer resistance to grazoprevir. Grazoprevir has a different resistance profile than the "second-wave, first-generation" drugs, which means that it has important characteristics for tackling the challenge of re-treating patients with direct-acting antivirals. Grazoprevir is currently being developed for use in interferon-free therapy in combination with a NS5A inhibitor, elbasvir (MK-8742).

2.1 Introduction

The era of direct-acting antivirals (DAAs) was ushered in by the approval of telaprevir (along with boceprevir outside of Japan) in 2011. Triple therapy with telaprevir, pegylated interferon, and ribavirin raised the response rate for genotype 1 hepatitis C, which was previously considered difficult to treat, up to 80 %. In addition, this combination treatment took only 24 weeks as opposed to 48 or 72 weeks for previous treatments. This led to the effectiveness of treatment for genotype 1 becoming comparable to treatment for genotype 2, which caused genotype 1 to lose its designation as "difficult to treat" (Fig. 2.1). However, since telaprevir causes adverse reactions such as characteristic skin reactions, renal dysfunction, and anemia, it had to be administered with great care [1]. During this time, a series of new protease inhibitors was approved following telaprevir, including simeprevir (September 2013), asunaprevir (July 2014), vaniprevir (September 2014), and paritaprevir (September 2015).

Telaprevir is a small linear molecule that inhibits protease activity by covalently binding deep within the active center of the NS3/NS4A protease. Simeprevir is a small macrocyclic molecule that binds noncovalently to the active center of the protease, acting somewhat like a lid. Development of a protease inhibitor with the code name BILN 2061 was started around the same time as the development of

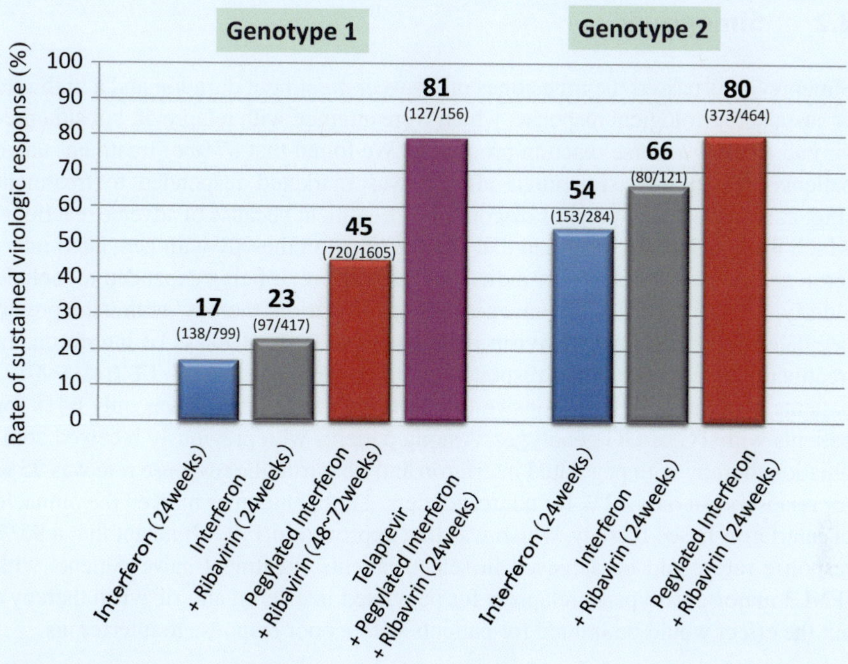

Fig. 2.1 Progress in antiviral therapy outcomes and the impact of telaprevir. Outcomes of antiviral therapy for hepatitis C at Osaka University and other institutions participating in the Osaka Liver Forum

telaprevir, but the development of that specific molecule was discontinued because of cardiac toxicity in large animals. However, an optimized version of the compound, simeprevir, was produced by modifying the structure of the molecule while retaining the same basic skeleton [2]. Asunaprevir and vaniprevir were also developed in a nearly identical fashion.

Since telaprevir and boceprevir are generally referred to as "first-generation" protease inhibitors, simeprevir, asunaprevir, vaniprevir, and paritaprevir are called either "second-generation" drugs or "first-generation, second-wave" drugs [3] because these drugs have characteristics that are different from those of telaprevir and boceprevir. One characteristic is a much lower incidence of clinical adverse reactions; these drugs do not exacerbate kidney injury or anemia like telaprevir and cause less severe skin reactions. They also have improved pharmacokinetic properties and a lower pill burden. For instance, whereas telaprevir had to be administered three times daily, second-generation drugs can be administered just once or twice daily. They have a low genetic barrier to resistance as do first-generation drugs but have a different resistance profile. The sites of resistance mutations for telaprevir are closer to the N-terminus of the NS3/NS4A protease (e.g., V36, T54, R155, and A156), whereas D168 is the most common site of resistance mutations for simeprevir, asunaprevir, vaniprevir, and paritaprevir.

2.2 Simeprevir

Simeprevir surpassed the milestones of 24-week treatment duration and a high rate of sustained virological response, which were attained with telaprevir, but simeprevir had a better adverse reaction profile [4]. We found that 87 % of treatment-naive patients treated with simeprevir after it was marketed responded to treatment (Fig. 2.2). Only 8 % of patients discontinued treatment because of adverse reactions, which is a much lower rate than that for combination therapy with pegylated interferon and ribavirin, despite the indication for simeprevir being expanded to include elderly adults [5]. The effectiveness of combination therapy with simeprevir, pegylated interferon, and ribavirin depends on the effectiveness of interferon. In treatment-naive patients, the response rate was 95 % for IFNL3 TT (rs8099917) patients, who are highly responsive to interferon therapy, but it was only 64 % for patients with TG or GG genotypes. Among patients who previously received combination therapy with pegylated interferon and ribavirin, the response rate was 95 % for relapsers but only 42 % for nonresponders. These outcomes marked the pinnacle of interferon-based therapy, which was first approved in 1992. It meant that a 95 % response rate could be achieved in select patients (treatment-naive patients with IFNL3 major genotype or relapsers for pegylated interferon and ribavirin therapy), but the effect would be limited for patients with a poor response to interferons.

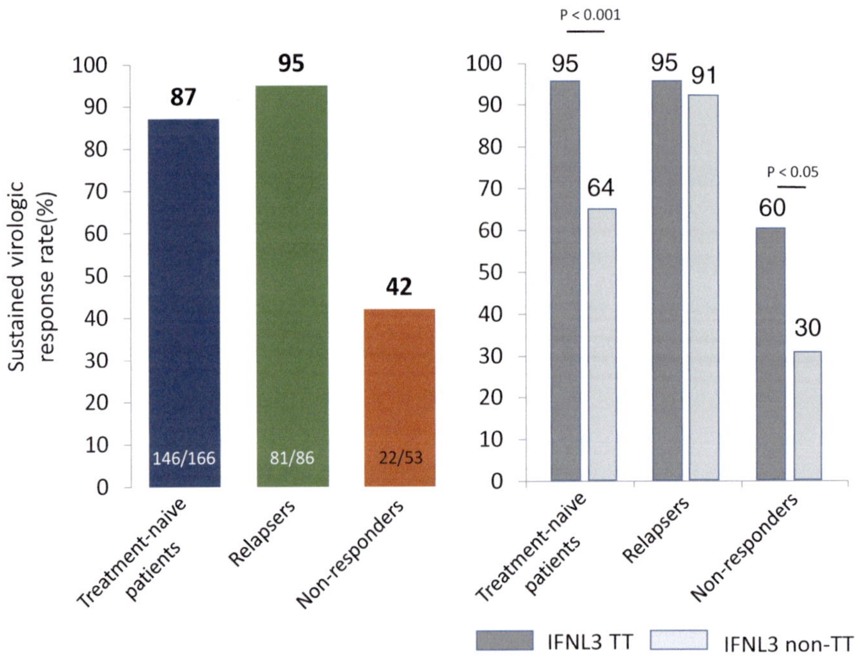

Fig. 2.2 Outcomes of three-drug therapy with simeprevir. Postmarket outcomes at Osaka University and other institutions participating in the Osaka Liver Forum

2.3 Asunaprevir

Unlike simeprevir and vaniprevir, which were developed for use in interferon-based therapy, asunaprevir was developed following a completely different strategy. Specifically, it was developed for interferon-free therapy combining DAAs with different mechanisms of action. Daclatasvir, a first-in-class drug that acts on NS5A, was selected to be coadministered with asunaprevir. Viral proteins cleaved by proteases arrange themselves on the membrane of the endoplasmic reticulum to form a membranous web, and viral genomic RNA is replicated within this spherical membrane structure. NS5A plays a vital role in the formation of this membranous web. Daclatasvir strongly suppresses viral proliferation by destroying the spherical membrane structure [6].

A 24-week regimen of asunaprevir and daclatasvir was the first interferon-free and ribavirin-free treatment regimen for hepatitis C to be approved in the world (July 2014). Sofosbuvir was approved in the United States at the end of 2013, but on the condition that it is combined with pegylated interferon and ribavirin for genotype 1 patients and with ribavirin for genotype 2 patients. After sofosbuvir was marketed, interferon-free therapy with simeprevir was sometimes used off-label for genotype 1 patients, while combination therapy with sofosbuvir and daclatasvir was also used in Europe. However, these regimens were never formally approved. The first interferon-free and ribavirin-free therapy approved outside of Japan was combination therapy with sofosbuvir plus ledipasvir, which was approved in the United States in November 2014 and in Europe in December 2014.

Clinical development of combination therapy with asunaprevir and daclatasvir was discontinued outside of Japan. This is because genotype 1a patients did not exhibit a sufficient response to this therapy [7]. However, it was approved in Japan because almost all Japanese genotype 1 patients have genotype 1b and clinical studies showed a response rate of 85 % in these patients [8]. The advantages of this treatment are that it has a comparable response rate for patients with compensated cirrhosis, a patient population who had previously been excluded from interferon-based DAA therapies, and that it is not affected by responsiveness to interferon therapy. In fact, the response rate for nonresponders to interferon therapy was 81 %, with no difference in outcomes regardless of stratification by IFNL3. The only baseline factor affecting response to treatment is NS5A mutations. The response rate was only around 40 % for patients with NS5A Y93H or L31M mutations. Substitutions capable of conferring resistance to daclatasvir are detected in about 20 % of Japanese patients prior to treatment. In contrast, as the asunaprevir resistance mutation D168V is only rarely detected at baseline, it was not considered a prognostic factor for response to combination therapy with asunaprevir and daclatasvir.

The main problem is that some patients who do not respond to this therapy (about 15 %) develop multiple resistance mutations within the NS3/NS4A protease domain and the NS5A replication complex domain. Typically, multidrug-resistant HCV strains with NS3/NS4A D168V, NS5A L31M/L31V, and NS5A Y93H mutations emerge. The NS3 mutation has poor fitness relative to wild type and becomes less frequent over time, but the NS5A mutation is known to persist for a long period of time. These kinds of resistance mutations are thought to limit options for DAA therapy later on.

It remains unclear whether combination therapy with sofosbuvir and ledipasvir, the next-generation oral pharmacotherapy (approved in Japan in July 2015), is effective in patients who failed to respond to combination therapy with asunaprevir and daclatasvir. We conducted a study in human hepatocyte chimeric mice to determine whether it is possible to employ combination therapy with ledipasvir and an NS5B polymerase nucleotide inhibitor to treat an HCV strain that has acquired resistance at three sites following combination therapy with asunaprevir and daclatasvir [9]. The resistant HCV was clearly more resistant to the ledipasvir and NS5B inhibitor therapy than wild-type HCV, and it was impossible to achieve a virological response after treatment in mice with resistant HCV. This strongly suggests that treatment with sofosbuvir and ledipasvir may be less effective than usual for nonresponders to combination therapy with asunaprevir and daclatasvir. It is also possible that the use of sofosbuvir when ledipasvir is ineffective might induce resistance to sofosbuvir (e.g., NS5B S282T). Therefore, the decision of whether to employ combination therapy with sofosbuvir and ledipasvir for nonresponders to asunaprevir and daclatasvir combination therapy must be considered with great care.

2.4 Vaniprevir

Approved in 2014, vaniprevir represents the last of the new interferon therapies [10]. Its pharmaceutical properties are very similar to those of simeprevir. It differs from simeprevir in that prior nonresponders receive the protease inhibitor for a total period of 24 weeks rather than 12 weeks. It yielded a relatively high response rate (62 %) in patients who did not respond to interferon therapy in clinical studies, which indicates that it may be effective for such patient groups. Its adverse reactions include upper gastrointestinal symptoms such as nausea and vomiting that occur particularly often during the early part of treatment. Its resistance profile is almost identical to that of simeprevir, but the R155 mutation sometimes occurs in addition to the D168 mutation in nonresponders.

2.5 Paritaprevir

Paritaprevir, the newest protease inhibitor, was approved in September 2015. It was designed to be coadministered with ritonavir, which delays drug metabolism and keeps the blood concentration of paritaprevir high enough that it only needs to be administered orally once daily. Paritaprevir was developed for use in interferon-free therapy in combination with the NS5A inhibitor ombitasvir. In Japanese clinical studies, it yielded a response rate of over 95 % when used in a 12-week dosing regimen. Response to treatment depends on the presence of NS5A resistance mutations at baseline. The response rate is 99 % for patients without the Y93H mutation at baseline and 87 % for patients with the mutation at baseline. This treatment is also known to be more effective for genotype 1b patients than genotype 1a patients. Therefore, when paritaprevir is used outside of Japan, it is used in a three-drug therapy with ombitasvir and the nonnucleoside polymerase inhibitor dasabuvir rather than the two-drug therapy with ombitasvir used in Japan [11].

2.6 Conclusion

Second-generation protease inhibitors have a similar resistance profile to first-generation drugs with the same Achilles' heel at the D168 residue of NS3/NS4A protease domain. In clinical practice, this creates a practical limitation in employing combination therapy with asunaprevir and daclatasvir to treat patients who did not respond to three-drug therapy with simeprevir. Patients previously treated with DAA therapy were excluded from treatment with asunaprevir and daclatasvir in the Japanese clinical trial. Patients previously treated with simeprevir have been treated with two-drug combination therapy with asunaprevir and daclatasvir after that treatment was marketed, but an increasing amount of research is showing a poor rate of response. This suggests that the D168 mutation acquired by patients who did not respond to simeprevir confers resistance to asunaprevir, thereby decreasing the effectiveness of this treatment. Asunaprevir therapy for patients treated with simeprevir is the first clinical attempt at re-treating patients previously treated with a DAA using a drug with the same resistance profile. This will continue to pose problems for re-treatment of patients who previously failed DAA combination therapy.

References

1. Oze T, et al. The prospective randomized study on telaprevir at 1500 or 2250 mg with pegylated interferon plus ribavirin in Japanese patients with HCV genotype 1. J Gastroenterol. 2015;50(3):313–22.
2. Lamarre D, et al. An NS3 protease inhibitor with antiviral effects in humans infected with hepatitis C virus. Nature. 2003;426(6963):186–9.
3. Sarrazin C, et al. Antiviral strategies in hepatitis C virus infection. J Hepatol. 2012;56 Suppl 1:S88–100.
4. Takehara T. Simeprevir for the treatment of chronic hepatitis C genotype 1 infection. Expert Rev Anti Infect Ther. 2014;12(8):909–17.
5. Berger C, et al. Daclatasvir-like inhibitors of NS5A block early biogenesis of hepatitis C virus-induced membranous replication factories, independent of RNA replication. Gastroenterology. 2014;147(5):1094–105.
6. Tahata Y, et al. The impact of an inosine triphosphate pyrophosphatase genotype on bilirubin increase in chronic hepatitis C patients treated with simeprevir, pegylated interferon plus ribavirin. J Gastroenterol. 2015;51:252–9.
7. Lok AS, et al. Preliminary study of two antiviral agents for hepatitis C genotype 1. N Engl J Med. 2012;366(3):216–24.

8. Kumada H, et al. Daclatasvir plus asunaprevir for chronic HCV genotype 1b infection. Hepatology. 2014;59(6):2083–91.
9. Kai Y, et al. Emergence of hepatitis C virus NS5A L31V plus Y93H variant upon treatment failure of daclatasvir and asunaprevir is relatively resistant to ledipasvir and NS5B polymerase nucleotide inhibitor GS-558093 in human hepatocyte chimeric mice. J Gastroenterol. 2015;50:1145–51.
10. Hayashi N, et al. Vaniprevir plus peginterferon alfa-2b and ribavirin in treatment-naive Japanese patients with hepatitis C virus genotype 1 infection: a randomized phase III study. J Gastroenterol. 2016;51:390–403.
11. Ferenci P, et al. ABT-450/r-ombitasvir and dasabuvir with or without ribavirin for HCV. N Engl J Med. 2014;370(21):1983–92.

The Efficacy of Daclatasvir Plus Asunaprevir Combination Therapy with Chronic Hepatitis

3

Fukiko Mitsui, C. Nelson Hayes, Fumitaka Suzuki, and Kazuaki Chayama

Abstract

In late years, therapeutic drugs for hepatitis C virus (HCV) have made impressive progress. In particular, the introduction of direct-acting antiviral agents (DAAs), which target viral proteins directly, has advanced the treatment of HCV with high sustained virological response (SVR) rates and few side effects. Despite their potency, monotherapy with a single agent is not effective due to rapid selection for strains harboring resistance-associated variants (RAVs). However, when DAAs with different mechanisms of action are used together, they present a high genetic barrier to resistance and have been shown to be

F. Mitsui (✉)
Department of Gastroenterology and Metabolism, Applied Life Sciences, Institute of Biomedical and Health Sciences, Hiroshima University, 1-2-3 Kasumi, Minami-ku, Hiroshima 734-8551, Japan
e-mail: fukifuki2448@yahoo.co.jp

C.N. Hayes
Department of Gastroenterology and Metabolism, Applied Life Sciences, Institute of Biomedical and Health Sciences, Hiroshima University, 1-2-3 Kasumi, Minami-ku, Hiroshima 734-8551, Japan

Liver Research Project Center, Hiroshima University, Hiroshima 734-8551, Japan

F. Suzuki
Department of Hepatology, Toranomon Hospital, 2-Chome, Toranomon, Minato, Tokyo 105-0001, Japan

K. Chayama
Department of Gastroenterology and Metabolism, Applied Life Sciences, Institute of Biomedical and Health Sciences, Hiroshima University, 1-2-3 Kasumi, Minami-ku, Hiroshima 734-8551, Japan

Liver Research Project Center, Hiroshima University, Hiroshima 734-8551, Japan

Laboratory for Digestive Diseases, Center for Genomic Medicine, RIKEN, Hiroshima 734-8551, Japan

K. Chayama (ed.), *Hepatitis C Virus Treatment*,
DOI 10.1007/978-981-10-2416-0_3

effective for treatment of patients with HCV genotype 1, including patients who are poor candidates for interferon therapy and patients who failed to respond to previous treatment attempts. In a phase 3 trial examining 24 weeks of dual DAA therapy with asunaprevir, an NS3 protease inhibitor, and daclatasvir, an NS5A replication complex inhibitor, in Japanese patients infected with HCV genotype 1 with a high HCV RNA titer, the SVR rate reached 81–91 %. Moreover, 94–100 % of patients without NS5A L31 and/or Y93 or NS3 D168 RAVs at baseline achieved SVR. The rate of adverse events was low, and this combination therapy was well tolerated. However, attention should be paid to AST/ALT elevation, pyrexia, and rash. Patients with cirrhosis are eligible for this therapy, but it is necessary to manage side effects carefully with consideration for the general condition of the patient. It is important to provide the best treatment currently available in the treatment of aging Japanese patients with hepatitis C to suppress progression of liver disease and reduce risk of hepatocellular carcinoma.

Keywords
Daclatasvir • Asunaprevir • IFN-free • Direct-acting antiviral agents (DAAs)

3.1 Introduction

Treatment of chronic hepatitis C virus (HCV) infection has improved considerably in recent years. The introduction of direct-acting antiviral agents (DAAs), which directly target viral proteins, has advanced the treatment of HCV with few side effects. Several DAA classes have become standard in the treatment of chronic HCV, including NS3 protease inhibitors, NS5A inhibitors, and NS5B polymerase inhibitors. Clinical trials of the 1st generation protease inhibitors telaprevir and boceprevir and a number of 2nd generation protease inhibitors, including simeprevir, vaniprevir, faldaprevir, asunaprevir, paritaprevir, and elbasvir, have demonstrated high sustained virological rates and resulted in few adverse events. Telaprevir, simeprevir, asunaprevir, and vaniprevir have already been approved in Japan. In addition, NS5A inhibitors (e.g., daclatasvir, ledipasvir, ombitasvir, grazoprevir) and NS5B polymerase inhibitors (e.g., sofosbuvir) have shown potent antiviral effects. DAA monotherapy is ineffective due to rapid emergence of antiviral resistance against individual agents [1, 2], but combination therapy involving two or more DAAs with nonoverlapping mechanism of action has been shown to have high safety and efficacy in treatment of patients with HCV genotype 1, including patients who have failed to respond to earlier therapy attempts. Because IFN- and ribavirin-free DAA therapies characteristically have few side effects, patients who cannot tolerate interferon or who are ineligible for treatment with IFN-based therapies, including the elderly and patients with compensated cirrhosis or comorbidities, can be successfully treated. In this article, I describe an all-oral combination therapy with asunaprevir (Sunvepra®) and daclatasvir (Daklinza®) which was approved in July 2014 for treatment of chronic hepatitis C and is used widely in Japan.

3.2 DAA Mechanisms of Action

3.2.1 Protease Inhibitors: Asunaprevir

The nonstructural (NS) protein 3-4A is a non-covalently bonded complex composed of the NS3 protease and the NS4A cofactor. NS3 is a 70 kDa multifunctional protein and includes a serine protease domain within the third part of the N-terminal region (aa 1–180) that is essential for cleaving the nonstructural proteins from the HCV polyprotein. Protease inhibitors inhibit virus proliferation by directly interfering with serine protease activity, thereby inhibiting the cleavage and maturation of viral proteins critical for replication of the viral genome and production of viral particles. Asunaprevir is a 2nd generation protease inhibitor with antiviral activity against HCV genotypes 1a, 1b, and 4.

3.2.2 NS5A Inhibitors: Daclatasvir

Nonstructural (NS) protein 5A is a dimeric phosphoprotein containing three domains and composed of 444 amino acids. NS5A contains the interferon sensitivity determining region (ISDR; aa2209–2248), which is associated with the response to IFN therapy, as well as the IFN/RBV resistance-determining region (IRRDR; aa2234–2379), which is associated with the response to IFN and RBV. Although the function of NS5A is not fully understood, it is considered to play an important role in the replication of viral RNA. It is thought to interact with the HCV core protein and participate in the formation of HCV particles.

NS5A inhibitors are low molecular weight agents expected to have a strongly inhibitive effect on viral replication. Daclatasvir is a first-in-class, NS5A replication complex inhibitor with potent pan-genotypic antiviral activity at picomolar quantities. Analysis of pharmacokinetics in healthy subjects and patients with HCV has shown that sufficient serum concentrations of daclatasvir can be maintained by daily internal use of 10 mg or more of daclatasvir [3].

3.3 Phase 3 Results of Daclatasvir Plus Asunaprevir Therapy in Japan

3.3.1 Treatment Results in Patients Who Did Not Respond to IFN-Based Therapy and Patients Who Are IFN-Ineligible/Intolerant

In a phase 2 study in Japan, 24 weeks of daclatasvir and asunaprevir combination therapy was well tolerated and resulted in a high SVR rate [4–6]. Therefore, a phase 3 study (AI447–026) was carried out that enrolled 87 prior nonresponders and 135 IFN-ineligible/intolerant patients [7]. The median ages in each group were 60 and 64 years, respectively, with male/female ratios of 39/48 and 38/97. The ratio of

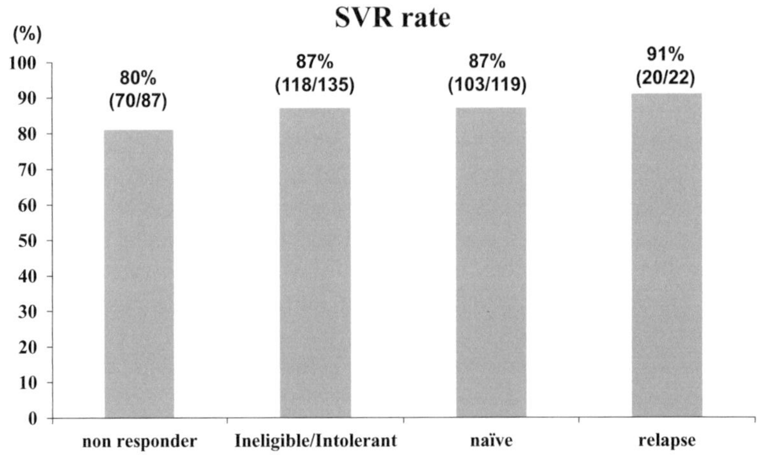

Fig. 3.1 Results of daclatasvir plus asunaprevir therapy for 24 weeks (Phase 3 trial in Japan; AI447-026, AI447-031)

IL28B rs12979860 CC/CT,TT genotypes was 16/71 and 94/41, respectively, with mean HCV RNA titers of 6.8 and 6.6 log IU/ml. Sustained virological response at 24 weeks after the end of treatment (SVR24) was achieved in 85 % of patients (Fig. 3.1). SVR was achieved by 91 % (20/22) of patients with cirrhosis. SVR rates were similar in both groups with respect to IL28B genotype, age, sex, HCV RNA titer at baseline, and cirrhosis. On the other hand, 11 patients (5.0 %) discontinued therapy; ten discontinued due to ALT and AST elevations, and one patient discontinued due to myasthenia gravis. The incidence of serious adverse events was low. The most common adverse events were nasopharyngitis, headache, and pyrexia. Grade 3/4 ALT elevations reversed rapidly after discontinuation of therapy. Eight of the ten patients who discontinued due to ALT/AST elevation subsequently achieved SVR.

Of the 34 patients who experienced virologic failure, 29 had resistance-associated variants against both daclatasvir (predominantly NS5A-L31M/V-Y93H) and asunaprevir (predominantly NS3-D168 variants) detected at the time of treatment failure. Twenty-two patients with virologic failure had pre-existing NS5A polymorphisms L31M/V and/or Y93H prior to treatment.

Of the 37 patients with L31M/V and/or Y93H substitutions at baseline, 4 out of 14 (29 %) prior nonresponders and 11 out of 23 (48 %) IFN-ineligible/intolerant patients achieved SVR. The NS3-D168E variant was present in two patients, one of whom achieved SVR. In all patients, 94 % (162/172) of patients without NS3-D168E, NS5A-L31M/V, and/or Y93H variants achieved SVR (Fig. 3.2).

Patients with ≥95 % compliance in dose and duration of treatment had an SVR24 rate of 93 % (179/193), compared with an SVR rate of 31 % (9/29) in patients who were <95 % compliant; 15 out of the 29 patients discontinued due to lack of efficacy.

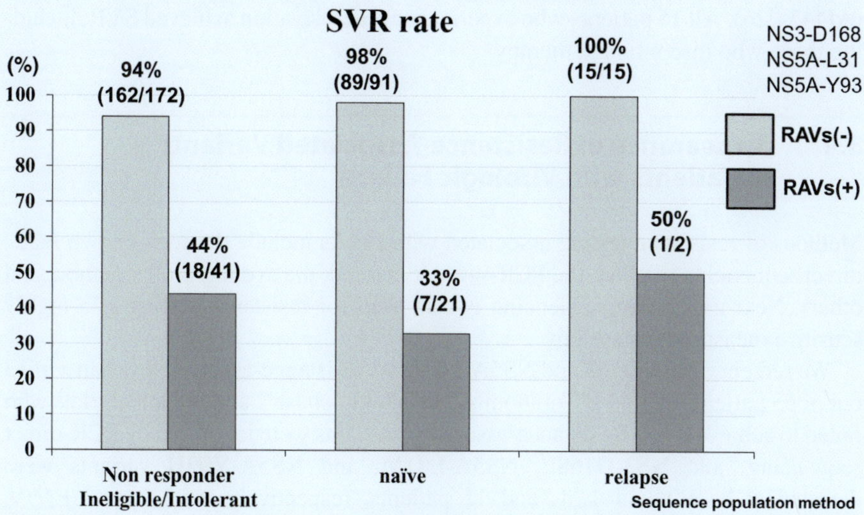

Fig. 3.2 Results of daclatasvir plus asunaprevir for 24 weeks with respect to resistance-associated variants (Phase 3 trial in Japan; AI447-026, AI447-031)

3.3.2 Treatment Results in Treatment-Naïve Patients and Prior Interferon Relapsers

Another phase 3 study (AI447-031) involving treatment-naïve patients and prior interferon relapsers was also carried out. In both groups, patients were infected with HCV genotype 1b and did not have cirrhosis. Treatment-naïve patients were randomly assigned to receive either daclatasvir plus asunaprevir dual therapy for 24 weeks or telaprevir plus peg-interferon/ribavirin triple therapy. Baseline characteristics of the daclatasvir plus asunaprevir group ($N = 119$) were as follows: median age 57 years, sex (male/female) 48/71, IL28 rs8099917 genotype (TT/TG+GG) 82/37, and mean HCV RNA 6.84 $\log_{10}$IU/ml. The SVR24 rate was 87 % among treatment-naïve patients who received daclatasvir plus asunaprevir therapy (Fig. 3.1), whereas the SVR24 rate was only 60 % in treatment-naïve patients treated with telaprevir plus peg-interferon/ribavirin, demonstrating the non-inferiority of daclatasvir plus asunaprevir compared to the standard of care.

Although 98 % (89/91) of patients without NS5A-Y93H and/or L31I/M variants achieved SVR, 33 % (7/21) of patients with these variants nonetheless achieved SVR (Fig. 3.2).

Twenty-two relapsers received 24 weeks of therapy with daclatasvir plus asunaprevir. Baseline characteristics of relapsers were as follows: median age 65 years, sex (male/female) 7/15, IL28 rs8099917 genotype (TT/TG+GG) 18/4, and mean HCV RNA 7.01 $\log_{10}$IU/ml. The overall SVR24 rate was 91 % (Fig. 3.1), although 100 % of patients without baseline NS5A-Y93H and/or L31I/M variants achieved SVR (Fig. 3.2). In most cases, adverse events were mild, as in the previous trial

(AI447-026). All 15 patients who experienced ALT elevation achieved SVR, including those who discontinued therapy.

3.4 Appearance of Resistance-Associated Variants in Patients with Virologic Failure

Methods of resistance testing associated with DAAs include the well-known PCR-direct sequencing method, the PCR-invader method, the cycling probe method, and others. Next-generation sequencing is also used for resistance testing as a highly sensitive measurement system.

We screened NS3-D168 and NS5A-L31/Y93 resistance-associated variants in 13 out of 63 patients treated at Toranomon Hospital in phase 2 and 3 clinical trials who failed to achieve SVR. Resistance-associated variants were identified by PCR-direct sequencing, and NS3-D168E, NS5A-L31M, and NS5A-Y93H variants were detected at baseline in 1, 1, and 11 patients, respectively. NS3-D168A/E/T/V, NS5A-L31I/M/V, and NS5A-Y93H variants were detected in 12, 12, and 13 patients, respectively, at the time of relapse. Two years after relapse, NS3-D168E/T, NS5A-L31I/M/V, and NS5A-Y93H variants were detected in 7, 11, and 13 patients, respectively. One patient with no NS3-D168 and NS5A-L31/Y93 resistant-associated variants at baseline had NS3-D168E, NS5A-L31I, and NS5A-Y93H variants at the time of relapse. However, 2 years after relapse, the NS3-D168E and NS5A-L31I variants had disappeared, and the frequency of the NS5A-Y93H variants reduced.

3.5 Treatment Results After Approval of the Therapy

We analyzed patients who had been treated with daclatasvir and asunaprevir combination therapy at our hospital between September 2014 and June 2015 following approval of the therapy for use in clinical practice. During this period 807 patients with the following characteristics were treated: sex (336 males and 471 females), median age 70 years (25–87), background liver disease (chronic hepatitis/compensated liver cirrhosis 483/324), history of treatment (treatment naïve/relapse after IFN treatment/ineligible or intolerant to IFN treatment/nonresponders to IFN 36/49/416/306). When patients who had participated in the phase 3 clinical trials were analyzed, virologic response rates at 4, 8, 12, 16, and 20 weeks under treatment and at the end (or cessation) of treatment were 83 %, 98 %, 99 %, 98 %, 99 %, and 96 %, respectively. The SVR4 rate (sustained HCV RNA response rate at 4 weeks after completion of treatment) was 89 %, and the SVR8 rate was 91 % (Fig. 3.3). The SVR8 rate was very high (97 %, 59/61) in patients who had no NS3-D168, NS5A-L31, and NS5A-Y93 variants at the start of treatment. On the other hand, the SVR rate was very low in patients who had been treated with daclatasvir and asunaprevir combination therapy less than 6 months after completing simeprevir plus peg-IFN/ribavirin therapy. Common adverse events were pyrexia, headache,

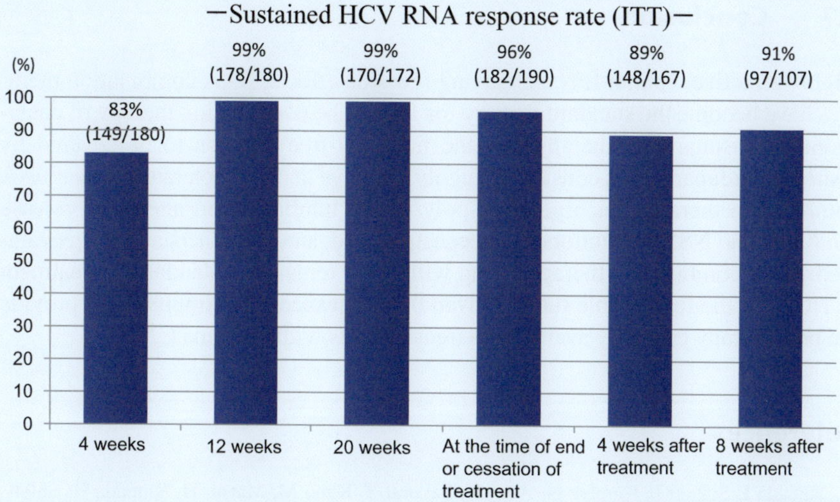

Inclusion criteria for the phase 3 trial: ①ALT 5 ULN or less, ②Total bilirubin <2.0mg/dl, ③Albumin ≥3.6g/dl , ④PT INR <1.7, ⑤WBC ≥1,500/mm³, ⑥Hb ≥8.5g/dl, ⑦Platelets ≥50,000/mm³, ⑧creatinine <1.8 ULN, ⑨ HCV RNA ≥5.0 logIU/ml, ⑩≤75 yrs, ⑪absence of hepatocellular carcinoma, ⑫no prior history of DAA therapy

Fig. 3.3 Daclatasvir plus asunaprevir combination therapy cases complying with phase 3 trial after approval (n = 194)

nasopharyngitis, skin rash, and AST/ALT elevation. The main reasons for discontinuation due to adverse events were AST/ALT elevation, viral breakthrough, skin disorders, pleural effusion/ascites, and pyrexia.

3.6 Indications and Cautions

Based on results from clinical trials as well as real-world post-marketing data, patients who do not have resistance-associated substitutions in NS5A-L31/Y93 or NS3-D168 at baseline are the best candidates for daclatasvir and asunaprevir combination therapy. As noted above, in a clinical trial in Japan, the SVR rate was 94–100 % in patients without resistance-associated variants (Fig. 3.2). However, the SVR rate was 40–50 % even in patients with pre-existing resistance-associated variants; it is possible that patients with advanced liver disease can still be treated with this therapy with adequate consideration.

The rate of the adverse events associated with daclatasvir and asunaprevir combination therapy is low, and the tolerability is good. However, it is necessary to consider rapid cessation of the therapy when AST/ALT levels reach grade 4 (i.e., >10 times the reference value). Although pyrexia is monitored while considering the dosage of the antifebrile, it is necessary to pay attention to concurrent infection. In patients with cirrhosis, it is necessary to manage the side effects carefully with consideration of the general condition of the patient.

3.7 Conclusion

Highly effective without IFN and having few side effects, DAA combination therapies have become the standard therapy for HCV. The combination therapy of daclatasvir and asunaprevir therapy was the first IFN-free regimen to be covered by insurance in Japan and is considered highly effective and well tolerated. Other DAA combination therapies incorporating polymerase inhibitors and alternate protease inhibitors and NS5A inhibitors have been approved, and still others are being evaluated. DAA combination therapy is and will likely remain the standard of treatment of HCV for the foreseeable future. Given these advances, it is important to provide the best therapy currently available to treat patients with hepatitis C.

References

1. Yamada I, Suzuki F, Kamiya N, Aoki K, Sakurai Y, Kano M, Matsui H, Kumada H. Safety, pharmacokinetics and resistant variants of telaprevir alone for 12 weeks in hepatitis c virus genotype 1b infection. J Viral Hepat. 2012;19:e112–9.
2. Toyota J, Ozeki I, Karino Y, Asahina Y, Izumi N, Takahashi S, Kawakami Y, Chayama K, Kamiya N, Aoki K, et al. Virological response and safety of 24-week telaprevir alone in japanese patients infected with hepatitis c virus subtype 1b. J Viral Hepat. 2013;20:167–73.
3. Gao M, Nettles RE, Belema M, Snyder LB, Nguyen VN, Fridell RA, Serrano-Wu MH, Langley DR, Sun JH, O'Boyle 2nd DR, et al. Chemical genetics strategy identifies an hcv ns5a inhibitor with a potent clinical effect. Nature. 2010;465:96–100.
4. Chayama K, Takahashi S, Toyota J, Karino Y, Ikeda K, Ishikawa H, Watanabe H, McPhee F, Hughes E, Kumada H. Dual therapy with the nonstructural protein 5a inhibitor, daclatasvir, and the nonstructural protein 3 protease inhibitor, asunaprevir, in hepatitis c virus genotype 1b-infected null responders. Hepatology. 2012;55:742–8.
5. Suzuki Y, Ikeda K, Suzuki F, Toyota J, Karino Y, Chayama K, Kawakami Y, Ishikawa H, Watanabe H, Hu W, et al. Dual oral therapy with daclatasvir and asunaprevir for patients with hcv genotype 1b infection and limited treatment options. J Hepatol. 2013;58:655–62.
6. Karino Y, Toyota J, Ikeda K, Suzuki F, Chayama K, Kawakami Y, Ishikawa H, Watanabe H, Hernandez D, Yu F, et al. Characterization of virologic escape in hepatitis c virus genotype 1b patients treated with the direct-acting antivirals daclatasvir and asunaprevir. J Hepatol. 2013;58:646–54.
7. Kumada H, Suzuki Y, Ikeda K, Toyota J, Karino Y, Chayama K, Kawakami Y, Ido A, Yamamoto K, Takaguchi K, et al. Daclatasvir plus asunaprevir for chronic hcv genotype 1b infection. Hepatology. 2014;59:2083–91.

4 Resistance-Associated Variants in the NS5A Region of HCV and Methods for the Detection

Satoshi Mochida

Abstract

NS5A is a phosphoprotein encoded by the nonstructural region of the HCV genome, in which nucleotide mutations occur frequently. The domain-1 of NS5A forms a dimer and acts as an RNA-binding groove for HCV replication. NS5A inhibitors such as daclatasvir inhibit the function of NS5A by binding to the groove. Thus, the antiviral effects of NS5A inhibitors may be attenuated when nucleotide mutations that are responsible for amino acid mutations on the surface of the groove appear. Genotype-1b HCV clones with amino acid mutations in the NS5A region that provoke resistance to NS5A inhibitors are present even in patients without previous antiviral therapies with direct-acting antivirals (DAAs). HCV clones with NS5A-Y93H mutations exist in about 15 % of treatment-naïve patients in Japan, and baseline profiles of HCV, such as NS5A-R30Q/H/L and NS5A-L31/M/V mutations as well as the NS5A-Y93H mutation, have been shown to be crucial for the outcome of antiviral therapies, including NS5A inhibitors. Resistance-associated variants (RAVs) in the NS5A regions can be identified using cycling-prove real-time PCR, an invader assay, and/or direct sequencing. The significance of NS5A-RAVs is currently under investigation in relation to host factors such as IFNL3-related gene polymorphisms and the response to antiviral therapies with interferon as well as DAAs.

Keywords

HCV • NS5A • Daclatasvir • Cycling probe

S. Mochida, MD, PhD (✉)
Department of Gastroenterology & Hepatology, Saitama Medical University, Saitama, Japan
e-mail: smochida@saitama-med.ac.jp

K. Chayama (ed.), *Hepatitis C Virus Treatment*,
DOI 10.1007/978-981-10-2416-0_4

4.1 Characteristics of HCV-NS5A and the Action Sites of NS5A Inhibitors

NS5A is a 447-amino acid phosphoprotein of 56 and 58 kDa, corresponding to basally phosphorylated (p56) and hyper-phosphorylated (p58) forms, respectively [1]. NS5A mainly consists of three domains. An amphipathic α-helix membrane anchor (aa5–aa25) exists in the N-terminal site of the domain-1 (aa28–aa213) and is responsible for the association of NS5A with the ER membrane [2]. Consequently, domain-1 of NS5A forms a dimer on the surface of the ER membrane and acts as an RNA-binding groove [3], which is crucial for RNA replication through the alteration of NS5B polymerase activity [4]. NS5A is also required for the assembly of viral particles through the action of domain-3 [5] and interacts with a variety of host cellular proteins [6].

HCV replicon-based high-throughput screens (HTS) have yielded several direct-acting antivirals (DAAs) that act as NS5A inhibitors. In Japan, daclatasvir, an NS5A inhibitor, was approved in July 2014, and the use of this agent in combination with asunaprevir, a second-generation NS3/4S protease inhibitor, has been approved for antiviral therapy in compensated patients with genotype-1b HCV infection since September 2014. Subsequently, ledipasvir, an NS5A inhibitor, was approved in 2015 as a compounding agent with sofosbuvir, a NS5B nucleotide polymerase inhibitor, for compensated patients with genotype-1b HCV infection. Also in 2015, ombitasvir, an NS5A inhibitor, was approved for use as a compounding agent with paritaprevir, a second-generation NS3/4S protease inhibitor, and ritonavir, a booster for the protease inhibitors. Thus, in Japan, NS5A inhibitors play a pivotal role in antiviral therapies for patients with genotype-1b HCV infection.

NS5A inhibitors suppress NS5A activity by binding to an RNA-binding groove in domain-1 [7]. Within the HCV genome, nucleotide mutations are especially abundant in the NS5A region. Thus, the antiviral effects of NS5A inhibitors may be attenuated when mutations result in non-synonymous amino acid substitutions located in the groove. HCV clones with such amino acid substitutions are present even in patients without previous exposure to DAAs [8], and the baseline profiles of HCV clones were shown to be crucial for determining the outcomes of antiviral therapies with NS5A inhibitors [9].

4.2 Assay Systems to Detect and Quantify NS5A-RAVs of Genotype-1b HCV

Among the various NS5A-RAVs showing resistance to NS5A inhibitors, HCV strains with NS5A-Y93H mutations were found most frequently among Japanese patients with genotype-1b HCV infection [8], and such strains showed a 24.44 times higher half maximal effective concentration (EC50) of daclatasvir, compared with that of wild-type HCV strains, based on experiments using a replicon system for genotype-1b HCV [10]. HCV strains with NS5A-L31M/V or NS5A-Y93C mutations were also detected in some patients with genotype-1b HCV infection, and the

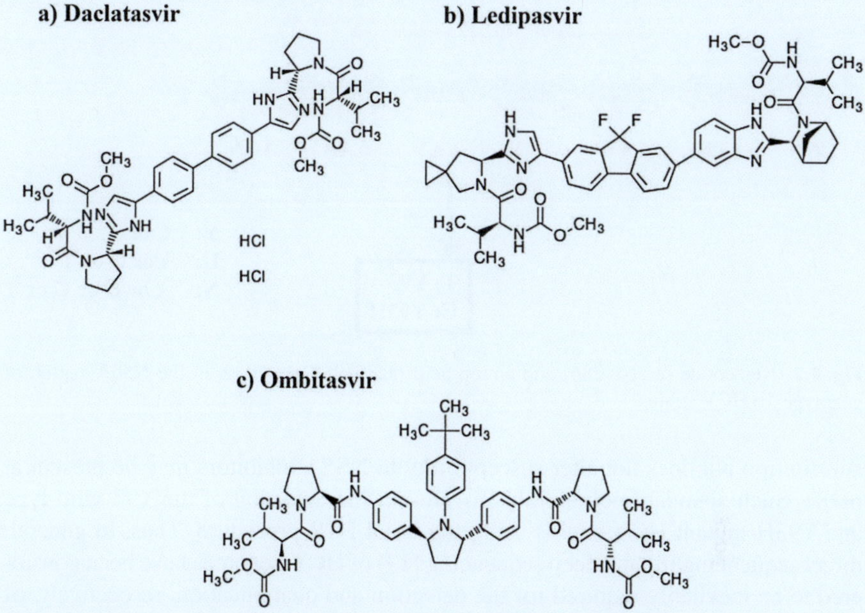

Fig. 4.1 Molecular structures of NS5A inhibitors approved for antiviral therapies for patients with genotype-1b HCV infection. (**a**) Daclatasvir, (**b**) Ledipasvir, (**c**) Ombitasvir

EC50 values of daclatasvir, compared to that in wild-type HCV strains, were 3.26, 27.89, and 3.32 times higher in the strains with NS5A-L31M, NS5A-L31V, and NS5A-Y93C mutations, respectively [10]. These NS5A polymorphisms were the most important factors for determining the antiviral efficacy of dual oral therapy with daclatasvir plus asunaprevir [9]; viral failure during or after therapy developed mostly in patients with pre-existing NS5A-Y93H mutant strains as a result of the emergence of double-mutant HCV strains with NS5A-Y93H and NS5A-L31M [11], which showed a 7,105.26-fold higher EC50 value for daclatasvir compared with that of wild-type HCV strains [10]. Moreover, it is noteworthy that quasispecies exist among HCV strains, especially regarding NS5A-Y93H mutations; the percent HCV-RNA levels of NS5A-Y93H mutant strains relative to the total HCV-RNA levels ranged in the sera between 1 % and 100 % in Japanese patients with genotype-1b HCV infection [12]. Thus, simple assay systems to detect and quantify NS5A-RAVs, especially NS5A-Y93H mutant HCV strains, are required to optimize the efficacies of antiviral therapies with DAAs, including NS5A inhibitors.

The genotype of nt277 in the NS5A region determines the phenotype of NS5A-Y93H mutations in HCV strains. As shown in Figs. 4.1 and 4.2, a number of synonymous nucleotide polymorphisms (N at nt267, nt276, nt282, nt285; H at nt270; and T at nt283, nt279) that do not alter the amino acid sequence are found at nt277, and a non-synonymous nucleotide polymorphism that provokes an amino acid

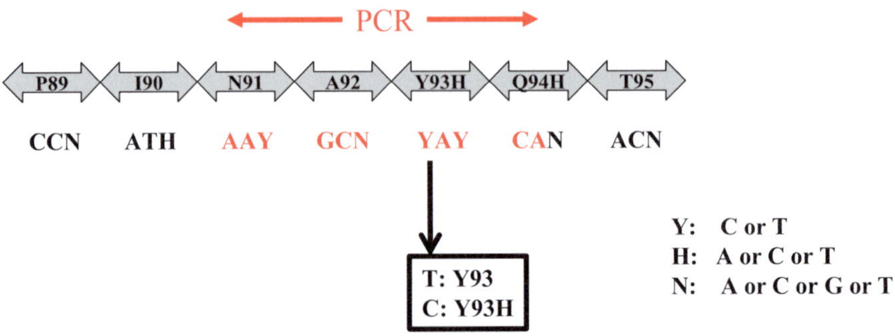

Fig. 4.2 Nucleotide (nt265–285) and amino acid (aa89–95) sequences in the NS5A region of genotype-1b HCV strains

substitution but does not alter susceptibility to NS5A inhibitors may be present at nt282. Such abundant polymorphisms prevent the detection of the Y93 wild-type and Y93H mutant HCV strains using the usual PCR procedure. Thus, in general, direct sequencing [8] and deep sequencing [13] of HCV genomes have been considered to be inevitably required for the detection and quantification, respectively, of NS5A-Y93H mutant HCV strains.

However, we overcame this problem through the adoption of cycling probe real-time PCR, in which a nucleoside in probes corresponding to nt278 was substituted for a nucleotide of the RNA [14]. In general, cycling probes are designed to hybridize the target portion (nt277) at the RNA portion, but the decision to shift the RNA portion enabled us to detect and quantify Y93 wild-type HCV strains and Y93H mutant HCV strains separately in almost all the patients with genotype-1b HCV infection [14]. Moreover, we established an assay system to determine NS5A-RAVs, including NS5A-L31M/V, as well as Y93H/C mutant HCV strains through combined cycling probe real-time PCR plus direct sequencing [12], and this system is commercially supplied to hepatologists in Japan in cooperation with SRL Inc. (Tokyo, Japan). In contrast, Kumada et al. established an assay system for NS5A-RAVs using the invader method [15], in which the amplification of nucleosides and nucleotides was not necessary through the adoption of structure-specific cleavage enzymes and a universal fluorescence resonance energy transfer (FRET) system. This system is also supplied commercially to hepatologists in cooperation with BML Inc. (Tokyo, Japan).

4.3 Significance of Amino Acid Mutations in the NS5A Region in Patients with HCV Infection

HCV strains with amino acid mutations in the NS3 region able to confer resistance to NS3/4A protease inhibitors are seldom found in patients who have not yet been exposed to DAA therapies. NS3-RAVs, however, develop frequently in patients who experience virologic failure during or after treatment with NS3/4A protease inhibitors, but these RAVS have been shown to disappear from the sera after therapy discontinuation in most patients [16]. In general, the proliferative capacity of strains harboring NS3-RAVs is lower than that of wild-type HCV strains, and wild-type HCV strains prevail as the major clone in the sera as well as in the liver, even when the serum RNA of such strains becomes undetectable during antiviral therapies with NS3/4A protease inhibitors. In contrast, NS5A-RAVs, such as Y93H/C or L31M/V mutations, are found in antiviral treatment-naïve patients. Previously, we used a combination of cycling probe real-time PCR and direct sequencing and reported that HCV strains with NS5A-Y93H/C and NS5A-L31M/V mutations were found in 87 patients (19.6 %) and 8 patients (1.8 %), respectively, out of 444 Japanese patients with genotype-1b HCV, in whom antiviral therapies with DAAs had not been performed [12]. Patients with HCV strains carrying NS5A-Y93H/C mutations included 54 patients (12.2 %) with mixed NS5A-Y93 wild-type and NS5A-Y93H mutant HCV strains and 33 patients (7.4 %) with exclusively NS5A-Y93H/C mutant HCV strains, including two patients with NS5A-Y93C mutant strains [12]. Patients carrying NS5A-L31M/V mutations included seven patients (1.7 %) with NS5A-L31M mutant strains and one patient (0.2 %) with an NS5A-L31V mutant strain; among the former, two patients (0.4 %) with double mutants (NS5A-L31M and NS5A-Y93H) were included [12]. The percentages of patients with NS5A-RAVs differed depending on the method used for detection; HCV strains with NS5A-Y93H mutation were shown to be found in 8.3 % and 30.9 % of Japanese patients with genotype-1b HCV infection using direct sequencing [8] and deep sequencing [13], respectively.

The NS5A polymorphisms in genotype-1b HCV were associated with the clinical features of patients, such as sex, serum AFP levels, presence of hepatocellular carcinoma, and IFNL3-related polymorphisms [12]. NS5A-Y93H/C mutant strains were found more frequently among women than among men, among patients without HCC and/or those with serum AFP levels less than 6.0 ng/mL than among patients with HCC and/or those with AFP levels of 6.0 mg/mL or more, and among those carrying the favorable IFNL3 single nucleotide polymorphism (SNP) than among those carrying an unfavorable allele. Multivariate analysis, however, identified IFNL3-related SNPs as the only independent factor associated with the presence of NS5A-Y93H/C mutations. We also demonstrated that the viral response against interferon may be superior in patients with NS5A-Y93H mutant HCV strains than in those with NS5A-Y93 wild-type strains through an evaluation in patients with mixed NS5A-Y93 wild-type and NS5A-Y93H mutant strains [12]. These observations suggest a close relationship between HCV and the genomic background of the host (Fig. 4.3); NS5A-Y93 wild-type HCV strains, which are insensitive to interferon, may be preferentially selected in patients carrying an unfavorable SNP allele in the IFNL3 gene, and endogenous interferon-λ expression might contribute to this phenomenon.

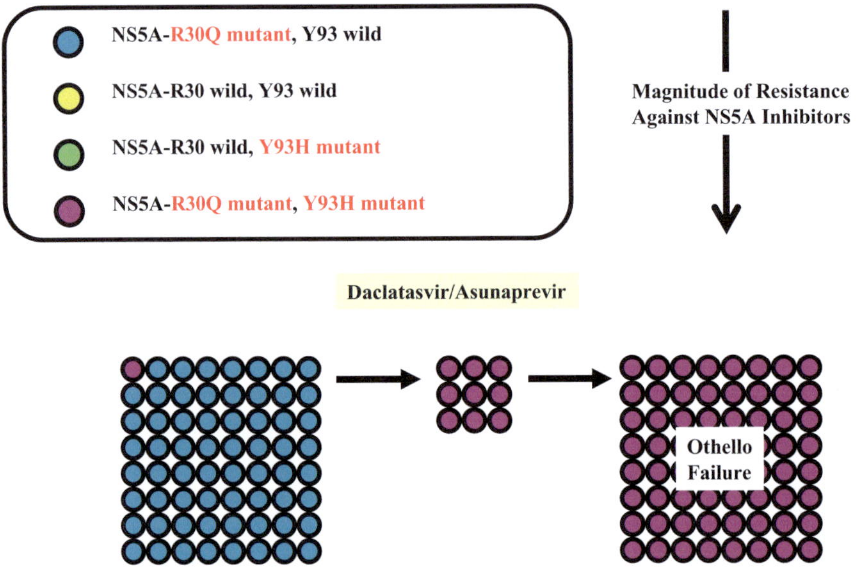

Fig. 4.3 Othello hypothesis involved in the development of virologic failure during and after dual oral therapy with daclatasvir plus asunaprevir in patients with HCV strains without NS5A-Y93H mutation

4.4 Therapeutic Efficacies of DAAs and NS5A Polymorphisms

NS5A-RAVs were shown to be crucial for the development of virologic failure during and after dual oral therapy with daclatasvir plus asunaprevir; SVR was obtained in 91.3 % of patients with wild-type NS5A-Y93 at baseline, while the SVR rate was 43.3 % among those with strains with pre-existing NS5A-Y93H mutations according to a clinical trial conducted in Japan [9]. In our institute, a total of 206 patients with genotype-1b HCV infection received dual oral therapy with daclatasvir plus asunaprevir between September 2014 and January 2015, and RAVs in the NS5A region at baseline and during/after therapy were evaluated using cycling probe real-time PCR, direct sequencing, and ultra-deep sequencing [17]. SVR12 was achieved in 180 patients (87 %); the rates were 95 % among patients without baseline NS5A-RAVs and 83 %, 59 %, and 77 % among those with HCV strains carrying NS5A-L31M, NS5A-Y93H/C, and NS5A-R31Q/H/L mutations, respectively. A multivariate analysis revealed baseline NS5A-R30Q/H/L and NS5A-Y93H mutations as significant factors associated with SVR12 [17]. HCV strains with NS5A-R30Q/H/L mutations were found in 26.0 % of the patients enrolled in our study and were shown to be a crucial factor associated with the antiviral efficacy of daclatasvir

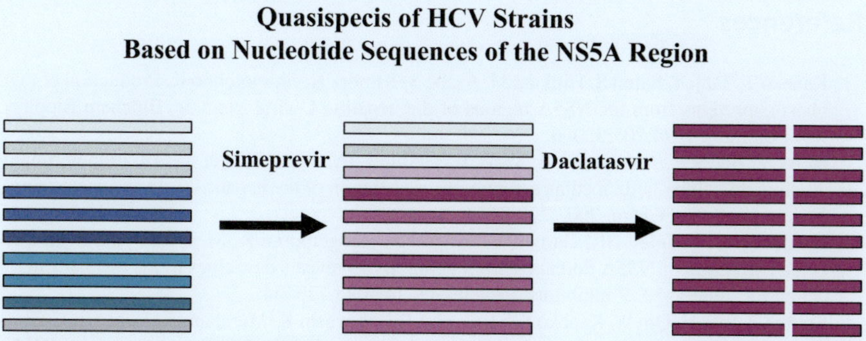

Fig. 4.4 Two-hit hypothesis involved in the development of rare NS5A-RAVS during and after dual oral therapy with daclatasvir plus asunaprevir in patients previously receiving triple therapy with simeprevir

plus asunaprevir. Although the EC50 of HCV strains with the NS5A-R30Q mutation against daclatasvir was reported to be 0.58 times that of wild-type HCV strains with the replicon system for genotype-1b HCV [10], that of HCV strains with NS5A-Y93H as well as NS5A-R30Q mutations against ombitasvir was 284 times greater than that of wild-type HCV strains, and the value was higher than that of HCV strains with NS5A-Y93H and NS5A L31M mutations (Interview Form of Viekirax, AbbVie Inc., North Chicago, IL). Thus, minor HCV strains with the NS5A-Y93H mutation as well as the NS5A-R30Q mutation might exist in the sera of patients with NS5A-R30Q strains, and these strains may contribute to virologic failure during and after dual oral therapy with daclatasvir plus asunaprevir. Thus, we proposed the "Othello hypothesis" [17] explaining the development of virologic failure in patients with HCV strains without the NS5A-Y93H mutation (Fig. 4.3).

In our study, virologic failure developed in all five patients who had been previously treated with simeprevir, a NS3/4A protease inhibitor, and rare RAVs, such as HCV strains with NS5A-29*del* and NS5A-32*del*, developed virologic failure during dual oral therapy with daclatasvir plus asunaprevir. Ultra-deep sequencing revealed that HCV strains with NS5A-29*del* or NS5A-32*del* were absent at baseline and emerged within 4 weeks of dual oral therapy among the strains appearing after simeprevir administration, suggesting that rare RAVs may develop in a two-hit manner, with simeprevir altering the quasispecies of HCV strains in the NS5A region, leading to the emergence of HCV strains with NS5A-29*del* and NS5A-32*del* during exposure to daclatasvir/asunaprevir [17]. These data prompted us to propose a two-hit hypothesis underlying the development of rare NS5A-RAVs during antiviral therapies with NS5A inhibitors (Fig. 4.4).

References

1. Kaneko T, Tanji Y, Satoh S, Hijikata M, Asabe S, Kimura K, Shimotohno K. Production of two phosphoproteins from the NS5A regions of the hepatitis C virus genome. Biochem Biophys Res Commun. 1994;205:320–6.
2. Brass V, Bieck E, Montserret R, Wolk B, Hellings JA, Blum HE, et al. An aminoterminal amphipathic alpha-helix mediates membrane association of the hepatitis C virus nonstructural protein 5A. J Biol Chem. 2002;277:8130–9.
3. Lambert SM, Langley DR, Garnett JA, Angell R, Hedgethorne K, Meanwell NA, et al. The crystal structure of NS5A domain 1 from genotype 1a reveals new clues to the mechanism of action for dimeric HCV inhibitors. Protein Sci. 2014;23:723–34.
4. Shirota Y, Luo H, Qin W, Kaneko S, Yamashita T, Kobayashi K, Murakami S. Hepatitis C virus (HCV) NS5A binds RNA-dependent RNA polymerase (RdRP) NS5B and modulates RNA-dependent RNA polymerase activity. J Biol Chem. 2002;277:11149–55.
5. Appel N, Zayas M, Miller S, Krijinse-Locker J, Schaller T, Fiebe P, et al. Essential role of domain III of nonstructural protein 5A for hepatitis C virus infectious particle assembly. PLoS Pathog. 2008;4(3):e1000035. doi:10.1371/journal/ppat.1000035.
6. Macdonald A, Harris M. Hepatitis C virus NS5A: tales of a promiscuous protein. J Gen Virol. 2004;85:2485–502.
7. Gao M, Nettles RE, Belema M, Snyder LB, Nguyen VN, Fridell RA, et al. Chemical genetics strategy identifies an HCV NS5A inhibitor with a potent clinical effect. Nature. 2010;465:96–100.
8. Suzuki F, Sezaki H, Akuta N, Suzuki Y, Seko Y, Kawamura Y, et al. Prevalence of hepatitis C virus variants resistant to NS3 protease inhibitors or the NS5A inhibitor (BMS-790052) in hepatitis patients with genotype 1b. J Clin Virol. 2012;54:352–4.
9. Kumada H, Suzuki Y, Ikeda K, Toyota J, Karino Y, Chayama K, et al. Daclatasvir plus asunaprevir for chronic HCV genotype 1b infection. Hepatology. 2014;59:2083–91.
10. Fridell RA, Qiu D, Wang C, Valera L, Gao M. Resistance analysis of the hepatitis C virus NS5A inhibitor BMS-790052 in an in vitro replicon system. Antimicrob Agents Chemother. 2010;54:3641–450.
11. Karino Y, Toyota J, Ikeda K, Suzuki F, Chayama K, Kawakami Y, et al. Characterization of virologic escape in hepatitis C virus genotype 1b patients treated with the direct-acting antivirals daclatasvir and asunaprevir. J Hepatol. 2013;58:646–54.
12. Uchida Y, Kouyama J, Naiki K, Sugawara K, Ando S, Nakao M, et al. Significance of variants associated with resistance to NS5A inhibitors in Japanese patients with genotype 1b hepatitis C virus infection as evaluated using cycling-probe real-time PCR combined with direct sequencing. J Gastroenterol. 2016;51:260–70.
13. Miura M, Maekawa S, Sato M, Komatsu N, Tsutsumi A, Takano S, et al. Deep sequencing analysis of variant resistant to the non-structural NS5A inhibitors in patients with genotype 1b hepatitis C virus infection. Hepatol Res. 2014;44:E360–7
14. Uchida Y, Kouyama J, Naiki K, Mochida S. A novel simple assay system to quantify the percent HCV-RNA levels of NS5A Y93H mutant strains and Y93 wild-type strains relative to the total HCV-RNA levels to determine the indication for antiviral therapy with NS5A inhibitors. PLoS One. 2014;9:e112647.
15. Yoshimi S, Ochi H, Murakami E, Uchida T, Kan H, Akamatsu S, et al. Rapid, sensitive, and accurate evaluation of drug resistant mutant (NS5A-Y93H) strain frequency in genotype 1b HCV by invader assay. PLoS One. 2015;10 (6):e0130022. doi:10.13771/journal.pone.0130022.
16. Akuta N, Suzuki F, Seko Y, Kawamura Y, Sezaki H, Suzuki Y, et al. Emergence of telaprevir-resistant variants detected by ultra-deep sequencing after triple therapies in patients infected with HCV genotype 1b. J Med Virol. 2013;85:1028–36.
17. Uchida Y, Kouyama J, Naiki K, Sugawara K, Inao M, Imai Y, et al. Development of rare RAVs that are extremely tolerant against NS5A inhibitors during daclatasvir/asunaprevir therapy via a two-hit mechanism. Hepatol Res 2016 Feb 15. doi: 10.1111/hepr.12673. [Epub ahead of print]

Detection of Antiviral Drug-Resistant Variants in Chronic Hepatitis C by Deep Sequencing

5

Shinya Maekawa, Mitsuaki Sato, and Nobuyuki Enomoto

Abstract

Due to the recent development of various novel small molecule compounds called direct-acting antiviral (DAA) agents and their dramatic antiviral potency against hepatitis C virus (HCV), more than 90 % of patients can now eradicate HCV with DAA therapy, a drastic improvement over the days of interferon therapy both in terms of efficacy and tolerability. On the other hand, the emergence of resistance-associated variants (RAVs) during DAA-based therapy has become a major problem. Namely, due to their high potency and specificity, there is a risk that patients treated with DAAs might develop resistance to one or more classes of DAA. Because of high cross-resistance among the small number of approved DAAs and the limited number of targets, it is also possible for HCV strains resistant to all current DAAs to emerge. Although it is possible to some extent to predict the emergence of RAVs through screening for the presence of RAVs prior to therapy using direct sequencing or other methods, precise and accurate prediction is still not possible because the presence of RAVs at baseline does not guarantee treatment failure nor does the lack of preexisting RAVs preclude the emergence of resistance through de novo mutation during treatment. In order to determine whether the emergence of RAVs is predictable based on the composition of HCV quasispecies prior to treatment, in this study deep sequencing was used to correlate the presence of RAVs at baseline with the emergence of RAVs during treatment in patients treated with interferon-based DAA triple therapy versus interferon-free DAA therapy.

S. Maekawa (✉) • M. Sato • N. Enomoto
First Department of Internal Medicine, Faculty of Medicine, University of Yamanashi,
1110, Shimokato, Chuo, Yamanashi 409-3898, Japan
e-mail: maekawa@yamanashi.ac.jp

K. Chayama (ed.), *Hepatitis C Virus Treatment*,
DOI 10.1007/978-981-10-2416-0_5

Keywords
Hepatitis C virus (HCV) • Direct-acting antivirals (DAAs) • Resistance-associated variants (RAVs) • Deep sequencing • Phylogenetic analysis

5.1 Introduction

Hepatitis C virus (HCV) is a major cause of chronic liver disease worldwide, progressing to liver cirrhosis over the long course of infection and leading to the development of hepatocellular carcinoma at a rate of 8 % per year among patients with cirrhosis [1]. On the other hand, due to the remarkable recent advances in the development of novel direct-acting antivirals (DAAs), the sustained viral response (SVR) rate, defined as undetectable HCV 6 months after the end of therapy, can now be achieved in most HCV-infected patients [2], and the rate of advancement to cirrhosis or hepatocellular carcinoma is expected to be significantly reduced. At the same time, while DAAs are distinguished for their high potency as well as their low frequency of serious adverse events (SAE), a major problem associated with DAA therapy is the emergence of resistance-associated variants (RAVs).

Most DAAs for HCV in clinical use or under development are low-molecular-weight compounds that inhibit viral enzymatic activity by directly binding to viral proteins. Due to the highly targeted nature of DAA characteristics, the appearance of RAVs is a predictable and perhaps inevitable problem since HCV could easily acquire adaptive mutations that facilitate escape from DAA binding due to the error-prone polymerase that lacks proofreading capabilities. In the era of DAA therapy, therefore, clinicians should understand how RAVs emerge during DAA therapy in order to prevent the development of resistance and obtaining the best treatment outcome.

In this chapter, the association between RAVs and the treatment response to DAA therapy is described briefly by introducing our original deep sequencing data.

5.2 Deep Sequencing Analysis of RAVs and the Aim of RAV Analysis

At present, DAAs in clinical use target the replication phase of HCV life cycle and are classified into three types of inhibitor: (1) NS3/4A protease inhibitors, (2) NS5A replication complex inhibitors, and (3) NS5B polymerase inhibitors. NS5B polymerase inhibitors are further classified into nucleoside inhibitors (NIs) and non-nucleoside inhibitors (NNIs). Variants at V36, T54, R155, A156, D168, and V170 of the NS3 protein are known RAV hot spots among NS3/4A protease inhibitors, while variants at L28, R30, L31, P58, and Y93 of the NS5A protein are RAV hot spots among NS5A inhibitors [3, 4]. NS5B S282 substitutions confer resistance to NS5B nucleoside inhibitors (NS5B NIs), while M495 (thumb1), L419, R422,

M423, I482 (thumb2), M414, Y448, G554, S556, and D559 (palm1 and palm2) confer resistance to NS5B non-nucleoside inhibitors (NS5B NNIs) [5]. Although RAVs generally appear after the initiation of DAA therapy, it has also been reported that a subset of RAVs, particularly within the NS5A protein, are present in a substantial number of DAA treatment-naïve patients [6]. DAA-induced RAVs might exhibit cross-resistance to other DAAs of the same class, and, in the event of non-SVR, accumulation of RAVs might contribute to development of HCV strains resistant to all current DAAs used in therapy.

It is known that HCV exists as mixed populations of closely related variant viruses called "quasispecies" [7] and that quasispecies composition changes dynamically upon exposure to anti-HCV therapy. Since RAVs are also considered to appear or disappear as quasispecies, detailed analysis of viral quasispecies using deep sequencing might help to determine how clinically relevant RAVs develop [8]. In this study, deep sequencing was performed in order to explore the dynamics of RAV quasispecies.

5.3 Interferon-Based DAA Therapy

5.3.1 TVR/PEG-IFN/RBV Combination Therapy

In Japan, the first clinically available DAAs were NS3/4A protease inhibitors (PIs) approved for use in combination with pegylated interferon (PEG-IFN) plus ribavirin (RBV) as an extension of the current standard of care interferon therapy. Telaprevir (TVR), a first-generation PI, was approved in 2011, and two second-generation PIs, simeprevir (SMV) and vaniprevir (VPV), were approved in 2013 and 2014, respectively [9, 10].

In order to determine how clinically relevant HCV variants develop under IFN-based TVR/PEG-IFN/RBV combination therapy, we undertook a deep sequencing study and analyzed quasispecies changes over time in 34 patients with TVR/PEG-IFN/RBV combination therapy.

At first, early changes in genetic complexity after initiation of the triple therapy were determined by calculating Shannon's entropy (Sn) and mutation frequency (Mf), indexes of genetic complexity, and the correlation between changes in genetic complexity and the treatment response or genotype of a SNP in the IL28B(IFNL3) locus were investigated. Deep sequencing of the viral NS3 region was performed in all 34 patients for two time points (baseline and 12 h after the start of therapy) to examine early changes in genetic complexity of viral quasispecies (Mf and Sn) after initiation of triple therapy (12 h) (Fig. 5.1) [11]. As shown in Fig. 5.1, genetic complexity tended to decrease in the SVR and IL28B TT groups, while no such differences were observed in the non-SVR and in IL28B TG/GG groups. Since IL28B is a host factor associated with the response to IFN and since the IL28B SNP is associated with the rate of SVR in triple therapy, the decrease in the genetic complexity observed in the SVR/IL28B TT group could result from differences in the IFN

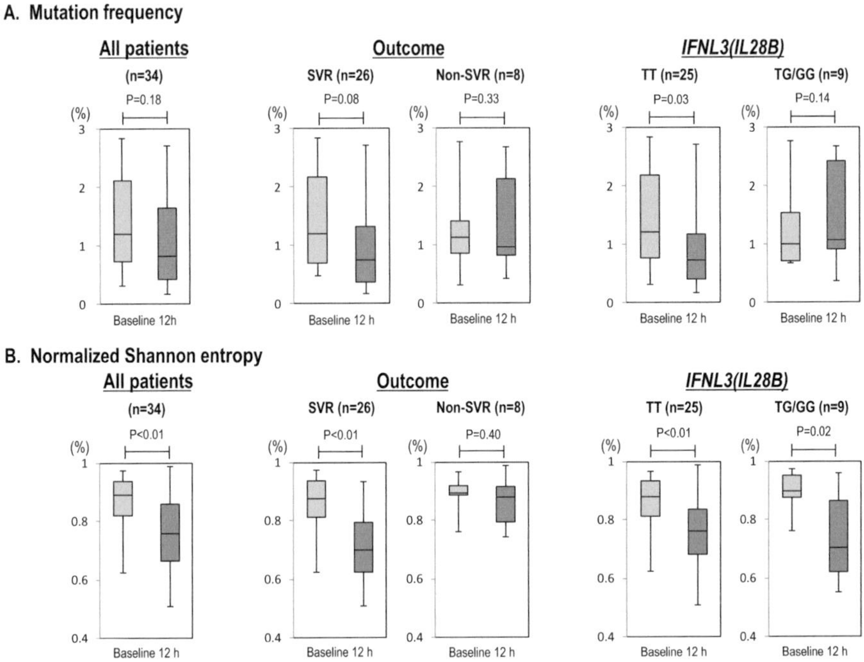

Fig. 5.1 Changes in the genetic complexity of the NS3 region 12 h after the introduction of TVR/PEG-IFN/RBV triple therapy

response between the two groups. It also suggested the possibility that changes in the genetic complexity as early as 12 h after the start of therapy might be used as a biomarker to predict SVR.

Next, we investigated changes in the viral sequence over time in the eight non-SVR patients (Table 5.1). A clinically resistant mutation was observed in five of the eight patients (62.5 %) during treatment and follow-up (V36C, 1; T54A, 2; A156F, 1; and A156S, 1). The same mutation was not observed at baseline even by deep sequencing in three of the five patients, but a T54A RAV was recognized at baseline as a minor population in two patients (Patients 3 and 7). The frequency of TVR-resistant RAVs was 98 % or more when the viral titer rebounded in all five patients, but the RAV frequency decreased in four patients (Patients 3, 4, 6, and 7) and was replaced by wild type as the dominant form during follow-up after the end of treatment.

Changes in the composition of HCV quasispecies over time were investigated in each of the non-SVR patients by constructing phylogenetic trees from all isolates obtained at three time points: baseline, re-elevation of the viral titer, and at the final observation. The isolates obtained at the re-elevation of the viral titer clustered differently from those at baseline (Fig. 5.2). Furthermore, isolates from the last observation point were distinct from those at baseline but seemed to be close to those obtained at the time of re-elevation (Figs. 5.2 and 5.3), demonstrating that the

Table 5.1 Telaprevir-resistant variants responsible for treatment failure in non-SVR patients and their time-dependent changes

Patient	Outcome	Previous response	IFNL3	Resistant variant				Variant rate (%)			
				V36	T54	R155	A156	Baseline	Re-elevation	Last observation (weeks after treatment)	
1	Discontinuation	Relapse	T/G								
2	Discontinuation	Relapse	T/T								
3	Discontinuation	Naïve	T/G		A			0.36	99.79	0.08	(26W)
4	Breakthrough	No response	T/T				F	0	98.19	0	(45W)
5	Relapse	Naïve	T/G								
6	Relapse	No response	T/T				S	0	99.48	1.27	(47W)
7	No response	Naïve	T/G		A			0.28	99.84	0.04	(34W)
8	No response	Naïve	T/G	C				0	98.11	96.05	(25W)

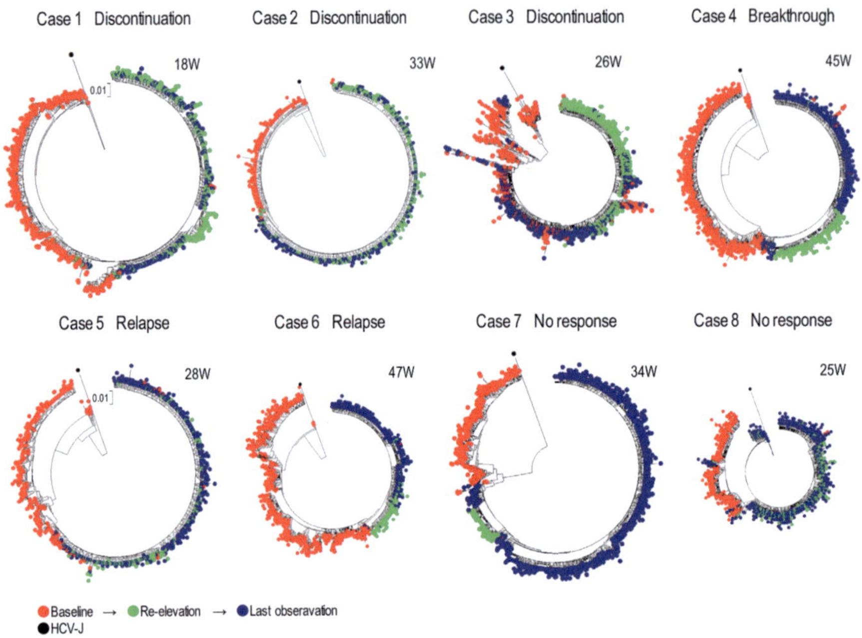

Fig. 5.2 Phylogenetic trees constructed for eight non-SVR patients using isolates at baseline, at re-elevation, and at the last observation. Numbers at the top right of each phylogenic tree indicate the number of weeks after the end of treatment

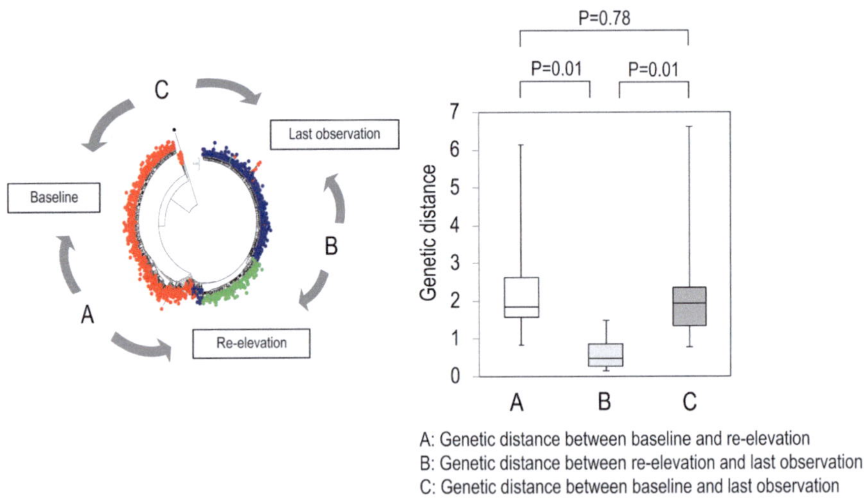

Fig. 5.3 Genetic distances in the NS3 region between baseline and re-elevation, between re-elevation and the last observation, and between the last observation and baseline in eight non-SVR patients. (**a**) Genetic distance between baseline and re-elevation. (**b**) Genetic distance between re-elevation and the last observation. (**c**) Genetic distance between baseline and the last observation

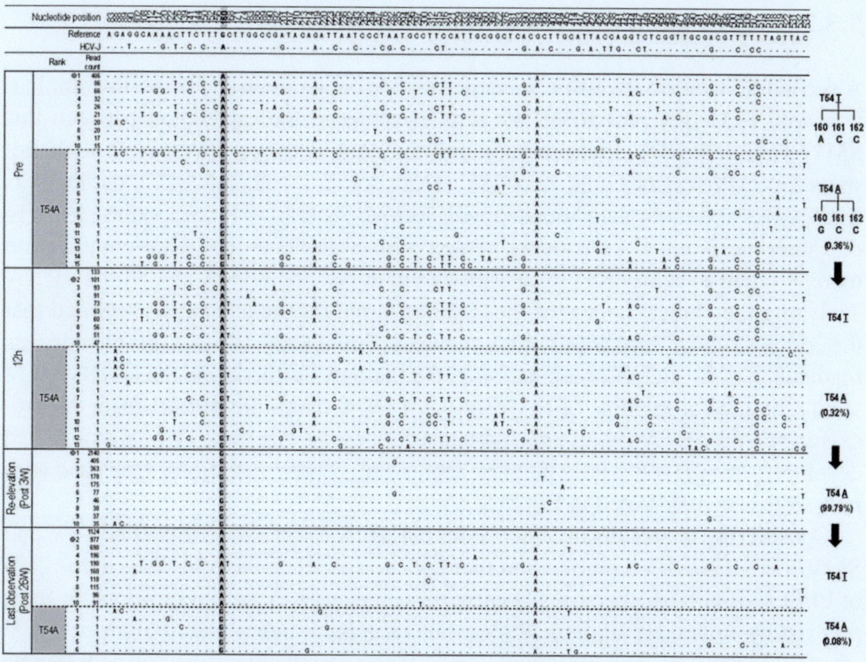

Fig. 5.4 Time-dependent changes in the top ten most populated isolates at baseline, 12 h, re-elevation, and last observation in Patient 3 who had RAV(T54A) at baseline in TVR/PEG-IFN/RBV therapy

changes in population composition induced by the triple therapy were unlikely to return to the pretreatment composition.

Lastly, to investigate which baseline populations developed TVR-resistant mutations, deep sequencing results over time were reanalyzed in patients in whom RAVs appeared (Patients 3, 4, 6, 7, and 8; Table 5.1). Among them, two patients (Patients 3 and 7) had T54A RAVs (0.36 % and 0.28 %, respectively) at baseline. In Patient 3, isolates with T54A accounted for 0.36 % of the population at baseline but had increased to 99.79 % at the time of re-elevation (Fig. 5.4). On the other hand, it was evident from comparison of their sequences that the T54A isolates at baseline were different from the major T54A isolate at the time of re-elevation. The baseline isolate most similar to the major T54A isolate at re-elevation was the wild-type isolate that existed as the dominant population before treatment, and we speculate that the T54A resistance mutation emerged from the wild-type isolate. Likewise, we speculate that the T54A isolate that became the dominant population after treatment was derived from wild type in Patient 7 (data not shown). These results suggest that the preexistence of RAVs does not strongly influence the probability of SVR in IFN-based triple therapy.

5.3.2 SMV/PEG-IFN/RBV Combination Therapy

A second-generation PI, SMV, was approved in Japan in 2013 for use in combination with PEG-IFN plus RBV. We also investigated the correlation between this SMV/ PEG-IFN/RBV triple therapy and the emergence of RAVs. NS3 RAV hot spots for SMV resistance include Q80, R155, A156, D168, and V170, but D168 is the RAV most frequently associated with second-generation PIs, including SMV. In our analysis, 3 out of 26 patients treated with SMV/PEG-IFN/RBV combination therapy failed to achieve SVR, and they each developed D168V. As with TVR, by analyzing the clinical course of these patients, phylogenetic analysis revealed that the composition of viral populations changed significantly and that this change was maintained even after discontinuation of the therapy. Furthermore, the D168V-HCV strains isolated at the time of re-elevation of HCV titer were distinct from the preexisting D168V-HCVs in all three patients, suggesting that these D168V strains had developed from D168 wild type after acquisition of new mutations. One case with non-SVR is demonstrated in Fig. 5.5.

Considering the results obtained from our analysis for TVR/PEG-IFN/RBV and SMV/PEG-IFN/RBV combination therapy, we speculate the following in the case of PI/PEG-IFN/RBV therapy: (1) antiviral-resistant HCV strains develop not from preexisting minor HCVs with RAVs, but emerge de novo from wild-type HCV after acquisition of new mutations, and (2) the composition of viral populations changes significantly after the development of clinically relevant RAVs, and this change is maintained even after the cessation of triple therapy and reversion to resistance variants to wild type. At present, the impact of the viral compositional change induced during PI/PEG-IFN/RBV therapy on the future of anti-DAA therapy is unknown. However, since this change in viral composition is speculated to be adaptive during exposure to triple therapy, it might weaken the response to future regimens based on PI/PEG-IFN/RBV, and further quasispecies study is needed.

5.4 Interferon-Free DAA Therapy

Recently, DAA therapies are evolving from "IFN-based therapies" to "IFN-free therapies" because of the efficacy, shorter treatment period, simplified administration, and low rate of SAEs. On the other hand, the problem of RAVs affecting the treatment outcome might be more serious in IFN-free therapies for the following reasons: (1) HCV cannot be eradicated by a single DAA agent alone at present, and, therefore, the combination of other classes of DAAs or other drugs such as ribavirin is needed. When two or three different classes of DAAs are combined, the risk in developing multiple DAA-RAVs is elevated. (2) In DAA combination therapy, NS5A inhibitors are frequently used as a backbone. However, it is reported that naturally occurring RAVs for NS5A inhibitors are more frequent than naturally occurring RAVs for PI inhibitors and NS5B inhibitors. (3) It has also been reported that the presence of NS5A RAVs at baseline significantly decreases the SVR rate.

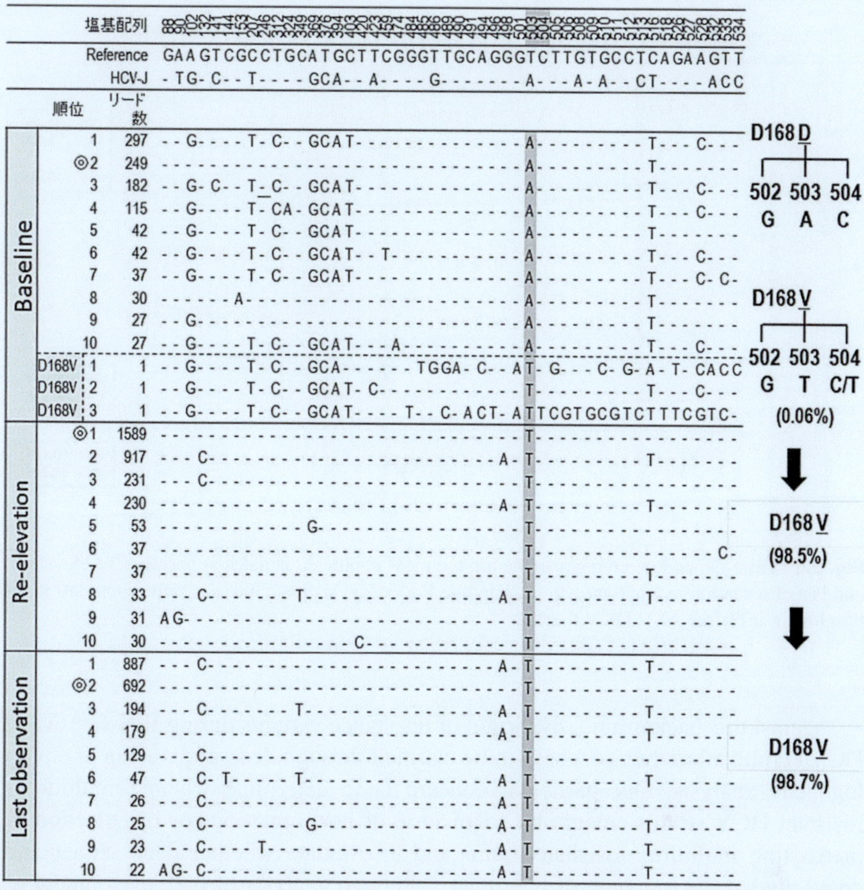

Fig. 5.5 Time-dependent changes in the top ten most populated isolates at baseline, re-elevation, and last observation in a patient who had RAV(D168V) at baseline in SMV/PEG-IFN/RBV therapy

(4) It was reported that RAVs for NS5A inhibitors persist for a long period of time because of their high replication fitness in comparison with wild type. (5) In IFN-containing regimens, the broad antiviral activity of IFN is robust in the presence of most DAA-related RAVs, and therefore, IFN may be useful in treating patients with RAVs.

In Japan, the first-approved interferon-free DAA regimen was the combination therapy of asunaprevir (ASV) and daclatasvir (DCV) in 2014 [12]. In this combination therapy, the presence of NS5A inhibitor resistance RAVs, particularly Y93H and L31M, is known to decrease the SVR rate [13] and may lead to the emergence of triple mutants (e.g., NS3-D168, NS5A-L31, and NS5A-Y93) at the time of treatment failure. HCV strains resistant to both ASV and DCV might develop further cross-resistance to other newer PIs and NS5A inhibitors.

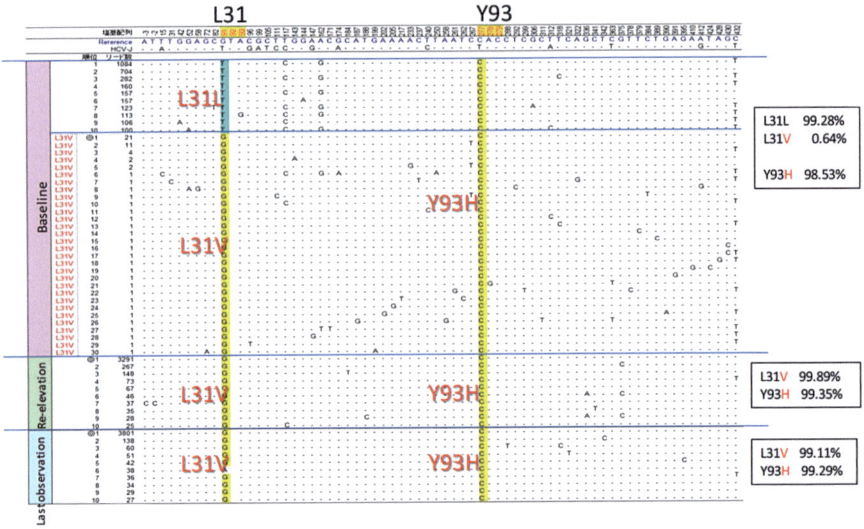

Fig. 5.6 Time-dependent changes in the top ten most populated isolates at baseline, re-elevation, and last observation in a patient who had double RAVs(L31V plus Y93H) as minor populations at baseline in IFN-free ASV/DCV therapy

Against this background, the origin of resistance variants during IFN-free ASV/DCV combination therapy needs to be clarified through deep sequencing and phylogenetic analysis. Our particular concern is to determine whether multidrug-resistant HCV strains emerge by acquisition of new mutations or by selection of preexisting multidrug-resistant strains and to evaluate whether deep sequencing could predict the outcome of treatment. Although analysis of the development of RAVs in ASV/DCV combination therapy is still under way, we encountered a patient for whom a clinically relevant multidrug-resistant HCV strain was considered to have emerged as the result of selection from a preexisting multidrug-resistant strain (Fig. 5.6). Namely, NS5A double mutants (L31M+Y93H) were observed prior to the start of therapy as a minor population, and this population was considered as the origin of the DAA-resistant HCV strain. Although further studies are needed, it is possible that selection for existing variants rather than the new mutations might play the dominant role in the emergence of multidrug-resistant RAVs during IFN-free DAA therapy.

5.5 Conclusions

In this chapter, we briefly described our deep sequencing study coupled with a phylogenetic analysis of RAVs in IFN-based and IFN-free DAA therapies. By applying this strategy, the origin and the role of RAVs in the clinical course could be more evident compared to conventional direct sequencing studies, and in the era of DAA therapy, understanding the role of RAVs is indispensable for clinical hepatologists to achieve the best treatment outcome.

References

1. Yoshida H, Shiratori Y, Moriyama M, Arakawa Y, Ide T, Sata M, Inoue O, et al. Interferon therapy reduces the risk for hepatocellular carcinoma: national surveillance program of cirrhotic and noncirrhotic patients with chronic hepatitis C in Japan. IHIT Study Group. Inhibition of Hepatocarcinogenesis by Interferon Therapy. Ann Intern Med. 1999;131:174–81.
2. Asselah T, Marcellin P. Direct acting antivirals for the treatment of chronic hepatitis C: one pill a day for tomorrow. Liver Int. 2012;32 Suppl 1:88–102.
3. Suzuki F, Sezaki H, Akuta N, Suzuki Y, Seko Y, Kawamura Y, Hosaka T, et al. Prevalence of hepatitis C virus variants resistant to NS3 protease inhibitors or the NS5A inhibitor (BMS-790052) in hepatitis patients with genotype 1b. J Clin Virol. 2012;54:352–4.
4. Sarrazin C, Kieffer TL, Bartels D, Hanzelka B, Muh U, Welker M, Wincheringer D, et al. Dynamic hepatitis C virus genotypic and phenotypic changes in patients treated with the protease inhibitor telaprevir. Gastroenterology. 2007;132:1767–77.
5. Di Maio VC, Cento V, Mirabelli C, Artese A, Costa G, Alcaro S, Perno CF, et al. Hepatitis C virus genetic variability and the presence of NS5B resistance-associated mutations as natural polymorphisms in selected genotypes could affect the response to NS5B inhibitors. Antimicrob Agents Chemother. 2014;58:2781–97.
6. Shindo H, Maekawa S, Komase K, Sueki R, Miura M, Kadokura M, Shindo K, et al. Characterization of naturally occurring protease inhibitor-resistance mutations in genotype 1b hepatitis C virus patients. Hepatol Int. 2012;6:482–90.
7. Martell M, Esteban JI, Quer J, Genesca J, Weiner A, Esteban R, Guardia J, et al. Hepatitis C virus (HCV) circulates as a population of different but closely related genomes: quasispecies nature of HCV genome distribution. J Virol. 1992;66:3225–9.
8. Nasu A, Marusawa H, Ueda Y, Nishijima N, Takahashi K, Osaki Y, Yamashita Y, et al. Genetic heterogeneity of hepatitis C virus in association with antiviral therapy determined by ultra-deep sequencing. PLoS One. 2011;6:e24907.
9. Hayashi N, Okanoue T, Tsubouchi H, Toyota J, Chayama K, Kumada H. Efficacy and safety of telaprevir, a new protease inhibitor, for difficult-to-treat patients with genotype 1 chronic hepatitis C. J Viral Hepat. 2012;19:e134–42.
10. Kumada H, Toyota J, Okanoue T, Chayama K, Tsubouchi H, Hayashi N. Telaprevir with peginterferon and ribavirin for treatment-naive patients chronically infected with HCV of genotype 1 in Japan. J Hepatol. 2012;56:78–84.
11. Sato M, Maekawa S, Komatsu N, Tatsumi A, Miura M, Muraoka M, Suzuki Y, et al. Deep sequencing and phylogenetic analysis of variants resistant to interferon-based protease inhibitor therapy in chronic hepatitis induced by genotype 1b hepatitis C virus. J Virol. 2015;89:6105–16.
12. Kumada H, Suzuki Y, Ikeda K, Toyota J, Karino Y, Chayama K, Kawakami Y, et al. Daclatasvir plus asunaprevir for chronic HCV genotype 1b infection. Hepatology. 2014;59:2083–91.
13. Fridell RA, Qiu D, Wang C, Valera L, Gao M. Resistance analysis of the hepatitis C virus NS5A inhibitor BMS-790052 in an in vitro replicon system. Antimicrob Agents Chemother. 2010;54:3641–50.

Incidence and Characteristics of Naturally Occurring Drug-Resistant Hepatitis C Virus Strains

6

Goki Suda, Jun Ito, Kenichi Morikawa, Koji Ogawa, and Naoya Sakamoto

Abstract

Advances in virology of the hepatitis C virus (HCV) have enabled rapid progress in developing direct-acting antivirals (DAAs) that directly target viral proteins. The use of these drugs has considerably improved treatment outcomes. However, the effects of these inhibitors are weakened by the presence of resistance-associated variants (RAVs). In addition, it is becoming clear that DAA-resistant variants exist in a certain proportion of cases that are naïve to DAAs. Genotype 1b variants that are resistant to NS3 protease inhibitors exist naturally, although they are uncommon. Variants resistant to nucleotide NS5B inhibitors are almost never observed in treatment-naïve patients. Conversely, the genotype 1b NS5A Y93H variant and the genotype 1a NS3 Q80K variant, which are resistant to NS5A or NS3/4 inhibitors, respectively, reportedly exist at relatively high frequency; hence, more care needs to be taken when administering DAA treatment.

Keywords

HCV • Direct-acting antivirals (DAAs) • Resistance-associated variants (RAVs) • NS5A inhibitor • Protease inhibitor

G. Suda, MD, PhD (✉) • J. Ito • K. Morikawa • K. Ogawa • N. Sakamoto
Department of Gastroenterology and Hepatology, Graduate School of Medicine, Hokkaido University, North 15, West 7, Kita-ku, Sapporo, Hokkaido 060-8638, Japan
e-mail: gsudgast@pop.med.hokudai.ac.jp

K. Chayama (ed.), *Hepatitis C Virus Treatment*,
DOI 10.1007/978-981-10-2416-0_6

6.1 Introduction

Hepatitis C virus (HCV) infects 170 million people worldwide and is a major pathogen causing liver cirrhosis and hepatocellular carcinoma [1, 2]. In Japan, there are over 1.5 million patients with chronic hepatitis C, many of whom are elderly and have increased risk of liver cancer. Because chronic hepatitis C is the primary cause of chronic liver disease and liver cancer, effective antiviral therapies are needed to reduce cancer-related deaths. For a number of years, combination therapy with interferon and ribavirin remained the standard of care therapy for chronic hepatitis C. However, the sustained virological response rate associated with this therapy in patients with HCV genotype 1 infection is insufficient at around 50 % [3]. In addition, the need for interferon-free therapies has become dire owing to the discovery of cases that are nonresponsive to interferon therapy, including those with an unfavorable SNP genotype in the IFNL3 locus [4].

In 1999, Lohmann et al. developed a way to culture HCV strains in vitro [5]. Along with advancements in the structural analyses of HCV proteins, this innovation led to rapid progress in developing direct-acting antivirals (DAAs), which directly target viral proteins. DAAs are broadly divided into three classes: those that inhibit the HCV NS3 protease, those that inhibit the HCV NS5A protein, and those that inhibit the NS5B polymerase. Combination therapies with these DAAs have reportedly achieved positive therapeutic effects in patients with an unfavorable (non-TT) genotype IFNL3 as well as in patients with liver cirrhosis and elderly patients, for whom interferon therapy is poorly tolerated [6, 7]. However, these inhibitors are reported to be less effective in the presence of resistance-associated variants (RAVs).

While these DAAs have extremely high antiviral activities, RAVs are known to appear at high frequencies when administered alone. As stated above, the therapeutic effects can decrease considerably when RAVs are present. In addition, it is becoming clear that DAA-resistant variants exist naturally in a certain proportion of cases that are naïve to DAAs. The NS5A Y93H and L31M/V variants, which are resistant to NS5A inhibitors, are reported to exist at relatively high frequencies in genotype 1b HCV cases. Similarly, naturally occurring DAA-resistant variants such as NS3 Q80K, which is known to attenuate the effects of simeprevir, exist at relatively high frequencies in patients with genotype 1a. In this chapter, we describe the frequency and resistance profiles of naturally occurring HCV variants and outline their viral characteristics.

6.2 Naturally Occurring Protease Inhibitor-Resistant Variants

DAAs that inhibit HCV protease were developed first, due to the early determination of the structure of the HCV protease. The HCV NS3/4A protein is known to function as a protease that enables HCV proteins to function by cleaving the nonstructural proteins from the translated HCV polyprotein. Therefore, inhibiting the

Table 6.1 Incidence of variants in the NS3 protease region

Position	Substitutions	Prevalence (%)
V36	V36A/M	0.3
T54	T54A/S	2.8–3.3
Q80	Q80K/L/R	0.7–2.2
R155	R155K/T/Q	0
A156	A156S/T/V	0.2
D168	D168A/E/V	0.7–1.2

Based on data from Refs. [8, 9]

activity of HCV protease can suppress viral replication. HCV protease inhibitors are divided into two groups based on their structure. One group is the first-generation protease inhibitors, which include telaprevir and boceprevir, which have a linear molecular structure with no branches. Succeeding these drugs were the second wave and second-generation drugs, such as simeprevir and vaniprevir, which have circular (macrocyclic) or branched molecular structures. There are RAVs with profiles unique to both first-generation and second-generation protease inhibitors.

Variants resistant to the first-generation protease inhibitor telaprevir include the V36, T54, R155, A156, and V170 variants. In a Japanese study using direct sequencing by Suzuki et al., V36A and T54S variants were detected in 0.3 % (1/307) and 3.3 % (10/307) of DAA-naïve cases, respectively [8]. An investigation of 493 cases by Itakura et al. found T54S and A156S in 2.8 % (14/493) and 0.2 % (1/493) of cases, respectively [9] (Table 6.1).

Known RAVs affecting the second-generation protease inhibitors simeprevir, vaniprevir, and asunaprevir include the Q80K/R, R155K, A156S/T, and D168V/A/E/T variants. Q80K is observed more frequently before treatment in genotype 1a cases. In a population sequencing study, these RAVs were found in about 20 % of genotype 1a cases, although they were relatively rare in genotype 1b cases, which are common in Japan [10]. In the Japanese study by Suzuki et al., Q80R was found in 0.7 % (2/307) of cases, no instances of R155K were found, and D168V/A/E/T was found in 0.7 % (2/307) of cases [8]. The investigation of 493 cases by Itakura et al. found Q80K in 2.2 % (11/493) and D168V/A/E/T in 1.2 % (12/493) of cases, indicating that these variants are relatively rare [9] (Table 6.1).

6.3 Naturally Occurring NS5A Inhibitor-Resistant Variants

The HCV NS5A protein is a phosphorylated protein formed from 450 amino acids. Its functions have not been fully clarified, but it is known that NS5A exists as a dimer, possesses an RNA-binding domain, and binds with HCV RNA. As it is a part of the HCV replication complex, it is known to be involved in viral replication, although it also has important functions in particle formation and virus release. NS5A inhibitors are thought to bind with NS5A domain 1. Since this is a relatively well-preserved structure even between genotypes, picomolar quantities of NS5A

Table 6.2 Incidence of variants in the NS5A region

Position	Substitutions	Prevalence (%)
L31	L31M	2.2–2.7
	L31V	0
Y93	Y93H	8.2–19

Based on data from Refs. [8, 9]

inhibitors successfully exert their effects against the HCV replicons of several genotypes. NS5A inhibitors include daclatasvir, which in Japan is approved for use in combination with the protease inhibitor asunaprevir; the recently approved ledipasvir, which is combined with the NS5B inhibitor sofosbuvir; and ombitasvir, which is expected to be approved for use in combination with protease inhibitors.

NS5A inhibitors, while having strong antiviral activity as described above, are problematic because of the high proportion of naturally occurring NS5A inhibitor-resistant variants. In a phase III trial for daclatasvir/asunaprevir combination therapy in Japan, NS5A variants conferring resistance (Y93H, L31M/V) were found in 31 out of 222 cases. The SVR rate of these 31 cases was poor, at 45 % (14/31) [6].

The existence of NS5A inhibitor-resistant variants in patients who have never been exposed to NS5A inhibitors has also been reported in Japan. In a direct sequencing study, Suzuki et al. observed Y93H variants in 8.2 % (24/294) and L31M in 2.7 % (8/294) of 294 cases. Furthermore, they observed the Y93H/L31 double mutant, which is known to be strongly resistant, in 0.3 % (1/294) of cases (Table 6.2). Itakura et al. reported Y93H in 19 % (78/410) and L31M in 2.2 % (9/410) of 410 cases. Miura et al. used next-generation sequencing to track NS5A inhibitor-resistant variants. They found Y93H in 30.9 % (34/110) and L31M/V/I in 11.8 % (13/110) of cases [11].

Surprisingly, the NS5A Y93H variant has also been reported to be associated with clinical factors. Miura et al. reported that Y93H was significantly more common in patients with the major IFNL3 SNP genotype. Itakura et al. also reported that Y93H was significantly more common when the IFNL3 SNP was the major type, as well as when platelet levels were low and the HCV viral load was high.

6.4 Naturally Occurring NS5B Inhibitor-Resistant Variants

The HCV NS5B protein functions as an RNA-dependent RNA polymerase and is essential for the replication of viral RNA. There are two classes of HCV NS5B inhibitors: nucleotide and non-nucleotide HCV NS5B inhibitors. Non-nucleotide inhibitors bind with NS5B catalyst domains to inhibit polymerase activity. Broadly, there are four types of non-nucleotide polymerase inhibitors that target different epitopes of the NS5B polymerase, and specific RAVs exist for each of them [12].

Nucleotide inhibitors inhibit replication of the viral genome when they are incorporated in the growing strand by the NS5B polymerase during viral RNA synthesis, causing chain termination. Thus, RAVs are less likely to arise for nucleotide

Table 6.3 Incidence of sofosbuvir-resistant variants in the NS5B region

Position	Substitutions	Prevalence (%)
S282	S282T	0–0.2

Based on data from Refs. [7, 14, 15]

polymerase inhibitors than for other DAAs, and these drugs possess antiviral activities against several HCV genotypes. Sofosbuvir is a nucleotide polymerase inhibitor that is currently being introduced in first-line therapies. In Japan it is being introduced clinically in combination with RBV to treat genotype 2 cases and is covered by health insurance to treat genotype 1-infected patients in combination with the NS5A inhibitor ledipasvir. A well-known sofosbuvir-resistant variant is S282T [13]. However, while this variant has been shown to dramatically decrease efficacy of the drug in vitro, it is reported to be very uncommon in nature. In a phase III trial for sofosbuvir as well as in database analysis, the S282T variant was not observed in a single case at baseline and is thought to be almost nonexistent in the natural environment [7, 14, 15] (Table 6.3). Thus, naturally occurring NS5B S282T variants are not expected to be a significant problem clinically.

6.5 Conclusion

As described above, drug-resistance variants are found to a certain extent in the natural environment. Depending on the treatment protocol, these RAVs can dramatically reduce the effectiveness of therapy. As such, the presence of RAVs before treatment should be investigated, and appropriate therapies should be selected. Further, it is becoming clear that RAVs appear at a high frequency in cases in which DAA therapy fails. Furthermore, new treatments designed to overcome RAVs that appear in the case of DAA treatment failure need to be investigated.

References

1. Mohd Hanafiah K, Groeger J, Flaxman AD, Wiersma ST. Global epidemiology of hepatitis C virus infection: new estimates of age-specific antibody to HCV seroprevalence. Hepatology. 2013;57:1333–42.
2. Seeff LB. Natural history of chronic hepatitis C. Hepatology. 2002;36:S35–46.
3. McHutchison JG, Lawitz EJ, Shiffman ML, Muir AJ, Galler GW, McCone J, Nyberg LM, et al. Peginterferon alfa-2b or alfa-2a with ribavirin for treatment of hepatitis C infection. N Engl J Med. 2009;361:580–93.
4. Tanaka Y, Nishida N, Sugiyama M, Kurosaki M, Matsuura K, Sakamoto N, Nakagawa M, et al. Genome-wide association of IL28B with response to pegylated interferon-alpha and ribavirin therapy for chronic hepatitis C. Nat Genet. 2009;41:1105–9.
5. Lohmann V, Korner F, Koch J, Herian U, Theilmann L, Bartenschlager R. Replication of subgenomic hepatitis C virus RNAs in a hepatoma cell line. Science. 1999;285:110–3.
6. Kumada H, Suzuki Y, Ikeda K, Toyota J, Karino Y, Chayama K, Kawakami Y, et al. Daclatasvir plus asunaprevir for chronic HCV genotype 1b infection. Hepatology. 2014;59:2083–91.

7. Mizokami M, Yokosuka O, Takehara T, Sakamoto N, Korenaga M, Mochizuki H, Nakane K, et al. Ledipasvir and sofosbuvir fixed-dose combination with and without ribavirin for 12 weeks in treatment-naive and previously treated Japanese patients with genotype 1 hepatitis C: an open-label, randomised, phase 3 trial. Lancet Infect Dis. 2015;15:645–53.
8. Suzuki F, Sezaki H, Akuta N, Suzuki Y, Seko Y, Kawamura Y, Hosaka T, et al. Prevalence of hepatitis C virus variants resistant to NS3 protease inhibitors or the NS5A inhibitor (BMS-790052) in hepatitis patients with genotype 1b. J Clin Virol. 2012;54:352–4.
9. Itakura J, Kurosaki M, Takada H, Nakakuki N, Matsuda S, Gondou K, Asano Y, et al. Naturally occurring, resistance-associated hepatitis C virus NS5A variants are linked to IL28B genotype and are sensitive to interferon-based therapy. Hepatol Res. 2015;45(10):E115–21.
10. Sarrazin C, Lathouwers E, Peeters M, Daems B, Buelens A, Witek J, Wyckmans Y, et al. Prevalence of the hepatitis C virus NS3 polymorphism Q80K in genotype 1 patients in the European region. Antiviral Res. 2015;116:10–6.
11. Miura M, Maekawa S, Sato M, Komatsu N, Tatsumi A, Takano S, Amemiya F, et al. Deep sequencing analysis of variants resistant to the non-structural 5A inhibitor daclatasvir in patients with genotype 1b hepatitis C virus infection. Hepatol Res. 2014;44:E360–7.
12. McCown MF, Rajyaguru S, Le Pogam S, Ali S, Jiang WR, Kang H, Symons J, et al. The hepatitis C virus replicon presents a higher barrier to resistance to nucleoside analogs than to non-nucleoside polymerase or protease inhibitors. Antimicrob Agents Chemother. 2008;52:1604–12.
13. Svarovskaia ES, Dvory-Sobol H, Parkin N, Hebner C, Gontcharova V, Martin R, Ouyang W, et al. Infrequent development of resistance in genotype 1–6 hepatitis C virus-infected subjects treated with sofosbuvir in phase 2 and 3 clinical trials. Clin Infect Dis. 2014;59:1666–74.
14. Omata M, Nishiguchi S, Ueno Y, Mochizuki H, Izumi N, Ikeda F, Toyoda H, et al. Sofosbuvir plus ribavirin in Japanese patients with chronic genotype 2 HCV infection: an open-label, phase 3 trial. J Viral Hepat. 2014;21:762–8.
15. Combet C, Garnier N, Charavay C, Grando D, Crisan D, Lopez J, Dehne-Garcia A, et al. euHCVdb: the European hepatitis C virus database. Nucleic Acids Res. 2007;35:D363–6.

Treatment Effects and Resistance-Associated Variants of Sofosbuvir Regimen for Japanese Patients with Chronic Hepatitis C Virus Genotypes 1 and 2

7

C. Nelson Hayes, Masaaki Korenaga, and Masashi Mizokami

Abstract

More than 40,000 people have already been treated in Japan since September 2014 when the first interferon-free all-oral therapy with daclatasvir (DCV), an NS5A inhibitor, and asunaprevir (ASV), an NS3/4A protease inhibitor (PI), was approved for treatment of chronic hepatitis C virus genotype 1. On the other hand, regimens containing the nucleic acid-type NS5B polymerase inhibitor sofosbuvir in combination with a PI, NS5A inhibitor, and/or ribavirin are now becoming standard throughout the world and achieving greater than 95 % sustained virological response (SVR) rates against multiple HCV genotypes. On March 23, 2015, sofosbuvir was approved for treatment of genotype 2 in Japan, priced at 61,700 yen per pill. Furthermore, Harvoni, a combination drug containing sofosbuvir co-formulated with the NS5A inhibitor ledipasvir, was approved in Japan on July 3, 2015, for treatment of genotype 1 with pricing set at 81,171 yen per pill. While the 6,800,000 yen price tag of 12 weeks of Harvoni is higher than that of DCV/ASV therapy, the therapy is more effective, and the one pill per day dosing and 12-week duration provides simpler and shorter therapy. Similarly, sofosbuvir is approved for and effective against both genotypes 1 and 2.

C.N. Hayes
Department of Gastroenterology and Metabolism, Applied Life Sciences, Institute of Biomedical and Health Sciences, Hiroshima University, 1-2-3 Kasumi, Minami-ku, Hiroshima 734-8551, Japan

Liver Research Project Center, Hiroshima University, Hiroshima 734-8551, Japan

M. Korenaga • M. Mizokami, MD, PhD (✉)
The Research Center for Hepatitis and Immunology, National Center for Global Health and Medicine, Tokyo, Japan
e-mail: mizokami0810@gmail.com

K. Chayama (ed.), *Hepatitis C Virus Treatment*,
DOI 10.1007/978-981-10-2416-0_7

This review discusses the mechanism and characteristics of sofosbuvir and summarizes results of phase III clinical trials for genotypes 1 and 2 and reports on the effects of antiviral resistance.

Keywords
Chronic hepatitis C • Sofosbuvir • Ledipasvir • Chain termination • Resistance-associated variants

Abbreviations

ASV	Asunaprevir
DAA	Direct-acting antiviral
DCV	Daclatasvir
HCV	Hepatitis C virus
LDV	Ledipasvir
PI	Protease inhibitor
RBV	Ribavirin
SOF	Sofosbuvir
SVR	Sustained virological response

7.1 The Mechanism of Action of Sofosbuvir and a Note of Caution

Direct-acting antiviral (DAA) agents for treatment of chronic hepatitis C virus (HCV) infection act by directly targeting viral structures. As noted in previous reports, monotherapy with a protease inhibitor or NS5A inhibitors rapidly induces antiviral resistance. Two types of agents that target the virally encoded NS5B-dependent RNA polymerase have also been developed, including nucleotide and non-nucleotide analogs. Sofosbuvir (SOF) is a nucleotide analog NS5B polymerase inhibitor that interferes with HCV replication by terminating strand RNA synthesis when incorporated as a defective substrate (chain termination). A key advantage of nucleotide analogs is that they are less vulnerable to resistance than non-nucleotide analogs.

7.2 Characteristics of Sofosbuvir

Gilead Sciences acquired the drug candidate PSI-7977 when it purchased the pharmaceutical company Pharmasset for $11 billion in 2011 [1]. In 2013, Gane et al. reported a 100 % (10/10) SVR rate among treatment-naive patients with HCV genotype 2 or 3 after 12 weeks of SOF plus ribavirin (RBV), and 60 % of patients achieved SVR with SOF monotherapy (Fig. 7.1). While patients with genotype 2 or

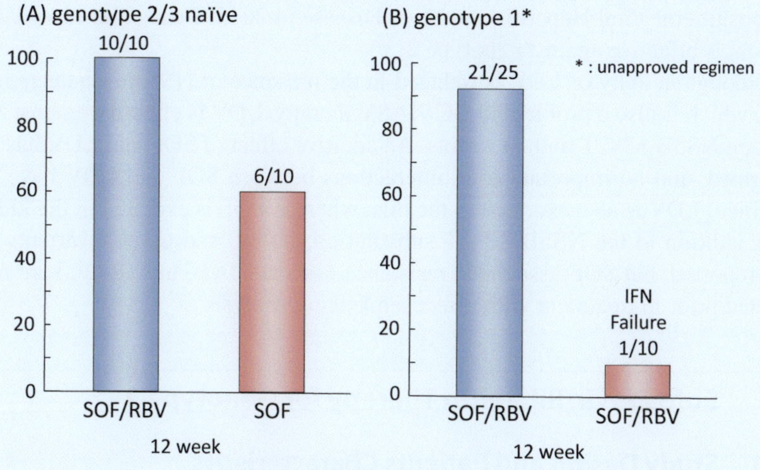

Fig. 7.1 SVR rate following sofosbuvir plus ribavirin therapy in (**a**) treatment-naive patients with genotype 2 or 3 (**b**) and in patients with genotype 1 [7]. Note that SOF+RBV therapy showed high response rates

3 can be successfully treated with SOF plus RBV, SVR rates in patients with genotype 1 are not as high. While 84 % (21/25) of genotype 1 patients achieved SVR with 12 weeks of SOF plus RBV therapy, only 10 % (1/10) of HCV genotype 1 patients who had failed prior IFN therapy achieved SVR. Therefore, treatment in combination with other DAAs, such as ledipasvir (LDV), is suggested for treatment of genotype 1. Nonetheless the success of these trials demonstrates the effectiveness of SOF against HCV.

SOF exhibits antiviral activity against multiple HCV genotypes, and the potential for drug interactions is low because SOF metabolism bypasses cytochrome P450. With over 300,000 people having already been treated with SOF since 2014, the drug's safety and tolerability have been well demonstrated, even among patients with renal dysfunction.

Development of many NS5B inhibitors has been abandoned due to cardiac toxicity. Although extremely rare in the case of SOF, bradycardia has been reported in nine patients treated in combination with amiodarone, and one patient experienced cardiac arrest following treatment with SOF in combination with a protease inhibitor. No causal relationship has been established, but caution should be exercised in clinical practice.

Harvoni is a co-formulated preparation of LDV and SOF for treatment of HCV genotype 1. Although the characteristics of LDV have been described in detail elsewhere, LDV shows stronger and more potent inhibitory activity against genotype 1 NS5A complex at picomolar concentrations compared to daclatasvir (DCV), another NS5A inhibitor already approved in Japan. Although in vitro studies suggest that the effect is somewhat reduced against genotypes 2 and 3, at the International

Symposium on Viral Hepatitis and Liver Disease in 2015, Gane et al. reported that Harvoni is effective against genotype 2.

Although activity of LDV is reduced in the presence of NS5A resistance mutations, which is also a problem in DCV/ASV therapy, LDV is effective against SOF-resistant NS5B S282T mutant strains. An additive effect of SOF and LDV has been confirmed, and no important drug interactions between SOF and LDV have been identified. LDV is also excreted in the bile, whereas SOF is excreted in the kidney.

In addition to the NS5B S282T substitution, L159F and C316N variants have been reported, but SOF-associated resistance-associated variants (RAVs) are rarely detected prior to treatment with direct sequencing analysis.

7.3 Sofosbuvir/Ribavirin Therapy for Genotype 2

7.3.1 Study Design and Patients Characteristics

A multicenter, randomized, open-label study was performed with the following inclusion criteria: (1) at least 20 years of age, (2) HCV RNA titer of at least 4 logIU/mL, (3) creatinine clearance of at least 1.0 mL/s [Cockcroft-Gault formula: men: (140−age)*weight/(72*serum creatinine); women: 0.85*(140−age)*weight/(72*serum creatinine)], and (4) platelet count of at least 50,000/mm^3. In total, 153 patients were enrolled, and all patients received 400 mg SOF once per day and RBV twice daily. RBV dosage was determined by body weight as follows: patients weighing less than 60 kg received 600 mg RBV, patients weighing 60–80 kg received 800 mg, and patients weighing more than 80 kg received 1000 mg. In addition, patients were classified according to prior treatment history. Ninety-three patients had no prior history of interferon treatment (naive), whereas 63 patients had failed to achieve SVR during prior treatment with IFN (treatment experienced). Patients were treated for 12 weeks with daily hospital visits for the first 6 weeks followed by hospital visits at 2-week intervals. The primary end point was attainment of SVR 12 weeks after the end of treatment (SVR12) (Fig. 7.2) [1, 2].

Patients with increased risk of HCC were also included in the study: 22 % of patients were over age 65, 46 % were male, and 11 % had liver cirrhosis. Cirrhosis was determined by liver biopsy or on the basis of a FibroScan score greater than 12.5 kPa within the previous 6 months. Patients with Child A decompensated cirrhosis or HCC were not included in the study.

7.3.2 Efficacy and Safety

All 153 patients successfully completed 12 weeks of SOF/RBV therapy without discontinuation of either drug, and all patients completed follow-up until 24 weeks after the end of treatment. The HCV RNA-negative (<25 IU/mL) rates during the course of therapy are shown in Fig. 7.3.

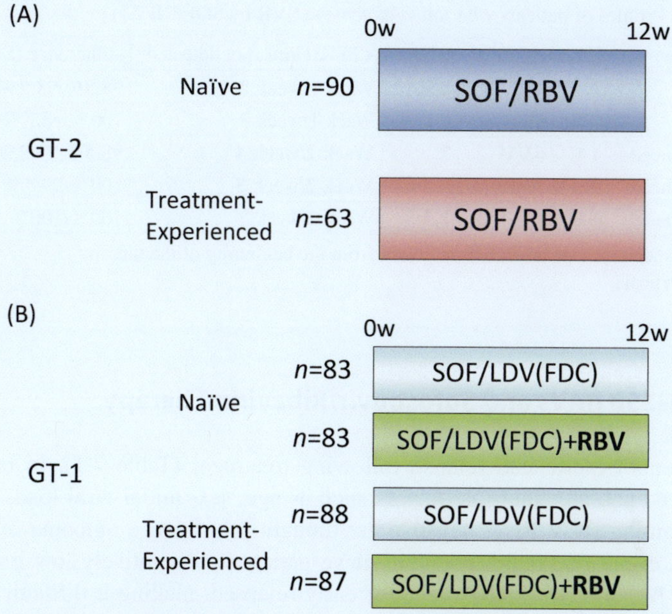

Fig. 7.2 Study design of SOF-based therapy for patients with genotype 2 (**a**) and those with genotype 1b (**b**). *FDC* fixed dose combination

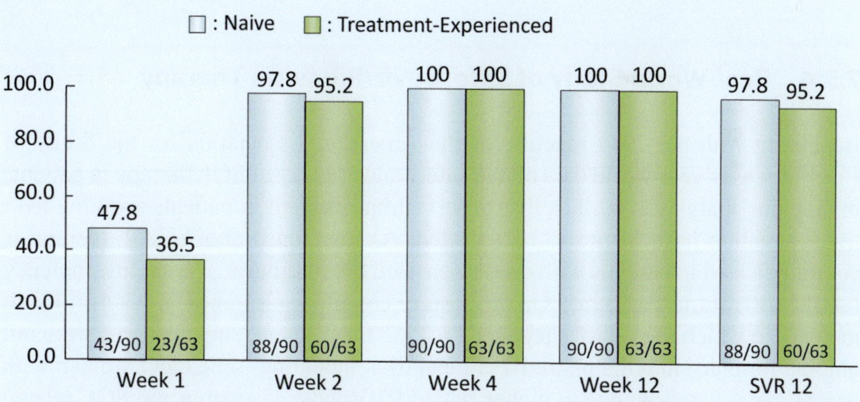

Fig. 7.3 HCV RNA-negative rate at key time points during and after therapy

All patients became HCV RNA negative by week 4, and all except five patients (two naive and three treatment experienced) remained negative until 12 weeks after the end of treatment (SVR12). It should be noted that all 148 patients who achieved SVR12 also achieved SVR24. About 5 % of patients experienced mild to moderate adverse events, including nasopharyngitis (30 %), headache (10 %), and RBV-induced anemia (10 %).

Table 7.1 Profiles of patients who failed to achieve SVR by SOF/RBV [1]

IFN history	LC	Age/sex	RNA	<25 KIU/mL/not detected	Adherence (SOF/RBV)
1. Naive	No	46/F	5.9	Week 1/week 2	98.5 %/98.5 %
2. Naive	No	**69**/F	6.1	Week 1/week 3	100 %/**93.7 %**
3. Experienced	**LC**	**63**/M	**7**	**Week 2/week 4**	98.5 %/98.2 %
4. Experienced	No	**70**/F	**6.8**	**Week 2/week 5**	100 %/99.4 %
5. Experienced	No	59/F	**7.3**	Week 2/week 3	100 %/100 %

HCV RNA-negative rate at each time points from the beginning of the therapy
LC liver cirrhosis

7.3.3 NS5B RAVs and Sofosbuvir/Ribavirin Therapy

Five patients experienced relapse following treatment (Table 7.1). In treatment-experienced patients, multiple factors such as age, sex, initial viral load, and time until becoming HCV RNA negative are thought to influence outcome of therapy. However, even one young treatment-naive patient with relatively low initial viral titer who became HCV RNA negative early relapsed, making it difficult to determine the cause. Although the NS5B S282T mutation reduces sensitivity to SOF, the presence of the S282T mutation before and after therapy has not been confirmed in these five patients, and it is unclear whether extension of therapy could improve the SVR in these patients.

7.3.4 Real-World Study of Sofosbuvir/Ribavirin Therapy

Beginning with the 2014 meeting of the American Association for the Study of Liver Diseases, real-world data on the safety and efficacy of SOF therapy in patients undergoing dialysis as well as liver reserve improvement in patients awaiting liver transplantation have begun to be reported. As mentioned above, SOF therapy is contraindicated in patients with decompensated liver cirrhosis or renal insufficiency (eGFR <30 mL/min/1.73 m^2). On the basis of creatinine clearance (Cockcroft-Gault method, in which the score is reduced by 15 % for women), some patients were not eligible for entry into the phase III clinical trial, including some older women with normal renal function, due to higher risk of RBV-induced anemia and SOF-related renal failure because the body weight of Japanese patients tends to be lower and patients tend to be older compared to western patients.

There was a 65-year-old male patient with HCV genotype 2b (7.8 log IU/mL HCV RNA) with compensated cirrhosis and type 2 diabetes with diabetic kidney disease (eGFR of 40 mL/min/1.73 m^2) who had failed to respond to prior PEG-IFN/RBV therapy. The patient was denied entry into the phase III clinical trial for genotype 2 due to anemia (Hb <12.0 g/dL) and renal dysfunction (creatinine clearance <1.0 mL/s). After FDA approval of SOF, the patient was able to begin self-financed

SOF/RBV therapy, but RBV dose reduction was necessary due to marked progression of anemia (ITPA: rs1127354 CC genotype). Furthermore, HCV RNA became undetectable at 8 weeks, showing a different clinical course than in the study patients, but the patient ultimately achieved SVR. Overseas guidelines recommend treatment of more than 16 weeks in GT-2 patients with cirrhosis, and the patient achieved SVR24 after a 4-week extension of the therapy to 16 weeks. In real-world data, in the case of a delay in HCV RNA decline due to RBV dose reduction, for example, it should be considered that 12 weeks of treatment may not be sufficient and longer treatment may benefit the patient.

7.4 Sofosbuvir/Ledipasvir Therapy for Genotype 1

7.4.1 Study Design and Patients Characteristics

The study design of the multicentre phase 3 clinical trial for SOF/LDV therapy for genotype 1 was similar to the genotype 2 study, except that selection criteria included HCV RNA ≥5 log IU/ml. The 341 enrolled patients were divided into four groups based on prior interferon treatment history (naive vs treatment experienced) and treatment with 400 mg SOF and 90 mg LDV with or without RBV (Fig. 7.2). SOF and LDV were administered for 12 weeks as a once daily co-formulated tablet. RBV dosage, treatment and follow-up timing, and study end points are the same as in the genotype 2 clinical trial.

About twice as many patients with cirrhosis (22 %) were enrolled compared to in the genotype 2 study. In addition, 33 % of the patients were at least 65 years old, and 28 patients over age 65 with cirrhosis were enrolled. There were 175 patients with prior interferon treatment history, about 60 % of whom had previously been treated with PEG-IFN/RBV and 25 % of whom had been treated with triple therapy that included an NS3/4A protease inhibitor. *As in the genotype 2 study, patients with Child A decompensated cirrhosis or HCC were ineligible for the study.*

7.4.2 Efficacy and Safety

The proportion of HCV-negative patients in each group and the time until HCV RNA became undetectable (<25 IU/mL) are shown in Table 7.2 and Fig. 7.4. All patients became HCV RNA negative by week 4 and remained negative until the end of treatment at week 12. There was no difference in the HCV RNA-negative rate between the RBV and RBV-free treatment arms.

Three hundred thirty-eight (99 %) patients achieved SVR12, all of whom also achieved SVR24. All three patients who failed to clear the virus were in the treatment-naive group treated with RBV. Two patients discontinued treatment due to (1) skin rash at day 6 and (2) and severe infection symptoms followed by cardiac arrest during week 8, respectively. Only one patient relapsed after completing the

Table 7.2 HCV RNA-negative rate at each time point during the therapy [2]

	IFN history (−)		IFN history (+)		
Time point and HCV RNA	SOF/LDV	SOF/LDV/RBV	SOF/LDV	SOF/LDV/RBV	Total
Negative rate (number of negative patients/number of measurements (%))	83	83	88	87	**341**
1 week	26/83 (31)	21/82 (25)	26/88 (30)	26/87 (22)	99/340 (29)
2 weeks	67/83 (81)	64/82 (78)	69/88 (78)	72/87 (83)	272/340 (79)
4 weeks	83/83 (100)	82/82 (100)	88/88 (100)	87/87 (100)	340/340 (100)
End of therapy	83/83 (100)	81/81 (100)	88/88 (100)	87/87 (100)	339/339 (100)

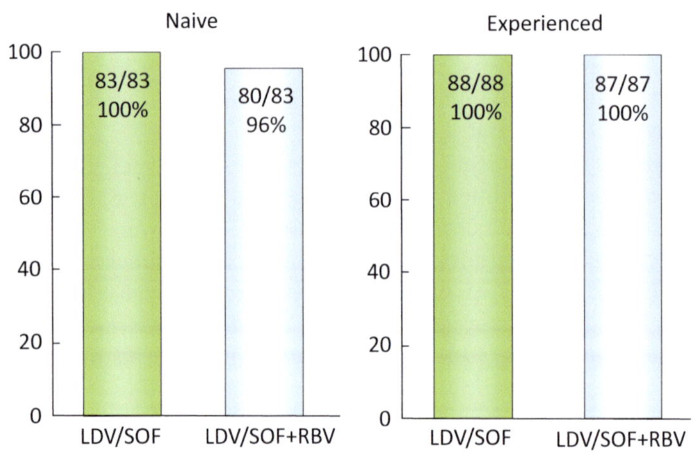

Fig. 7.4 SVR12 rate of patients in a phase III clinical trial of LDV/SOF. Patients who had never been treated with IFN-based therapy (naive, *left*) and patients with a prior history of IFN therapy (experienced, *right*) in total, 338/341 (99 %) patients achieved SVR12 [2]

12-week therapy. Adverse events were similar to that in the genotype 2 study and included nasopharyngitis, anemia, and headache. In addition, a 71-year-old male with cirrhosis in the treatment-experienced RBV arm experienced myocardial infarction 9 days after the end of therapy.

For these reasons, the use of RBV with LDV/SOF therapy is not approved in Japan.

7.4.3 NS5B Resistance-Associated Variants and Sofosbuvir/Ledipasvir Therapy

Figure 7.5 shows the effect of substitutions in the NS5A region on the outcome of LDV/SOF therapy. It should be noted that the presence of drug-resistant variants was determined using the Illumina MiSeq platform using a minimum frequency of 1 % as a cutoff. No resistance mutations were detected in 265 patients, 99 % of whom achieved SVR, except for two patients who discontinued therapy early. Pretreatment substitutions in the NS5A region were present in 76 patients (22 %), 58 of whom harbored NS5A Y93H mutations, but all patients achieved SVR except for one patient who relapsed. The presence of single NS5A substitutions does not appear to affect outcome of therapy as it does in DCV/ASV therapy.

Although the Y93H mutation was detected at a frequency of >99 % in one patient who relapsed, a similar pattern was detected in ten other patients who did not. Analysis of next-generation sequencing data is complicated by differences among sequencing hardware and methods that can lead to differences in reported values among institutions. In addition, at the time of relapse (4 weeks after the end of treatment), no mutations in the NS5A or NS5B regions other than NS5A Y93H could be detected. As in the case with genotype 2, extension of therapy may help to eliminate HCV in such patients.

7.4.4 Sofosbuvir/Ledipasvir Therapy in the Real World

Although SOF-based regimens are on the way to becoming the standard of care, IFN-free therapy with DCV/ASV arrived on the scene in Japan first. Although many patients successfully achieved SVR with this therapy, a downside of early adoption of this NS5A inhibitor-containing therapy is the high frequency of NS5A L31/Y93 double mutations [3] and P32 deletions [4, 5] among the several thousand patients who failed to respond to DCV/ASV therapy. Re-treatment of these patients with an NS5A inhibitor is likely to be less effective, presenting a major challenge for re-treatment with Harvoni.

Although one patient with a NS5A L31I+Y93H double mutation achieved SVR in the phase III clinical trial in Japan (Fig. 7.5), NS5A L31M/V+Y93H strains are strongly resistant to NS5A inhibitors in vitro, and the therapeutic effect in humans is unclear. In fact, while about 1 % of patients had NS5A L31+Y93 double mutants prior to oral DAA therapy, the emergent NS5A L31+Y93 substitutions can be detected even by direct sequencing in non-SVR DCV/ASV patients.

Strains from patients who have experienced DCV/ASV treatment failure due to the presence of resistance-associated variants have been reported to be suppressed in vitro with the addition of DCV/SOF or LDV/SOF (Harvoni) [6]. However, as described above, DCV and LDV share the same mechanism of action, and LDV is ineffective against DCV-resistant strains. Therefore, such treatment is effectively equivalent to SOF monotherapy, and the therapeutic effect is likely to be insufficient.

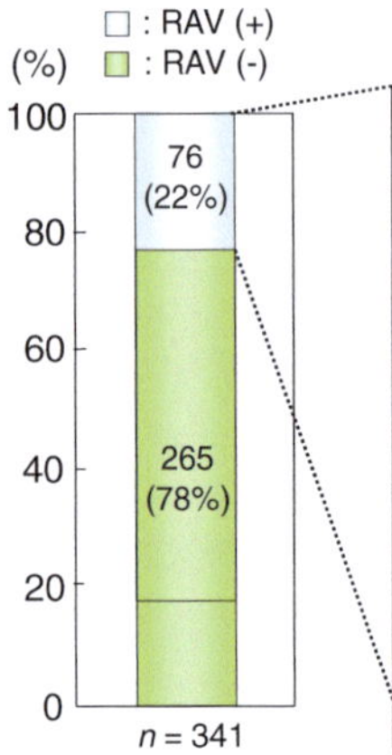

	Base line RAVs	genotype	Numberof patients	SVR24%(n/n)
Single RAV	L31M	1b	9	100(9/9)
	L31I	1b	1	100(1/1)
	L31F	1b	1	100(1/1)
	L31V	1b	1	100(1/1)
	Y93H	**1b**	**58**	**98(57/58)**
	Q30R	1a	5	100(1/1)
Multiple RAV	**L31I, Y93H**	1b	1	100(1/1)
	Y93S, Y93N, Y93H	1b	1	100(1/1)
	Y93F, Y93H	1b	1	100(2/2)
	L31I, Y93N, Y93C	1a	2	100(1/1)
		total	76	99(75/76)

Fig. 7.5 Outcome of therapy with respect to frequencies of single and multiple resistant-associated variants (RAVs) at baseline

At the 2015 meeting of the European Society of Hepatology, Lawits et al. reported results of re-treatment with 24 weeks of Harvoni for patients who failed to achieve SVR after 8 or 12 weeks of Harvoni therapy. The SVR rate for patients with NS5A resistance variants prior to therapy was 60 % (18/30). In addition, *emergent NS5B resistance mutations (S282T, L159F) were detected in 33 % of patients.* Even if a variety of novel oral antiviral agents are developed in the future, the emergence of multidrug-resistant strains should be avoided as much as possible, and caution should be used in re-treatment of patients who experience DAA treatment failure.

7.5 Conclusion and Future Perspectives

This review described the results of the first SOF phase III trial conducted in Japan as well as real-world treatment effects of SOF. The 100 % SVR rate of ribavirin-free LDV/SOF therapy for genotype 1 is surprising, especially since more than 30 % of the patients were aged 65 or older and more than 20 % of the patients had cirrhosis. Achieving a high therapeutic effect without IFN or ribavirin following a shorter 12-week therapy even in elderly patients is indeed very good news.

On the other hand, although the majority of patients respond to the therapy, 12 weeks of therapy is insufficient to clear the virus for some patients. As with patients who failed to respond to DCV/ASV, re-treatment in the case of DAA failure is an urgent challenge in order to suppress liver carcinogenesis and improve prognosis. Remaining challenges also include improved treatment options for (1) patients with genotypes other than genotypes 1 and 2, (2) patients with renal failure or who are undergoing dialysis, (3) patients with decompensated cirrhosis, and (4) genotype 2 patients who are intolerant of ribavirin.

As we are entering an era of high rates of treatment, failure to clear the virus must be avoided as much as possible.

References

1. Omata M, Nishiguchi S, Ueno Y, Mochizuki H, Izumi N, Ikeda F, et al. Sofosbuvir plus ribavirin in Japanese patients with chronic genotype 2 HCV infection: an open-label, phase 3 trial. J Viral Hepat. 2014;21(11):762–8.
2. Mizokami M, Yokosuka O, Takehara T, Sakamoto N, Korenaga M, Mochizuki H, et al. Ledipasvir and sofosbuvir fixed-dose combination with and without ribavirin for 12 weeks in treatment-naive and previously treated Japanese patients with genotype 1 hepatitis C: an open-label, randomised, phase 3 trial. Lancet Infect Dis. 2015;15(6):645–53.
3. Kumada H, Suzuki Y, Ikeda K, Toyota J, Karino Y, Chayama K, et al. Daclatasvir plus asunaprevir for chronic HCV genotype 1b infection. Hepatology. 2014;59(6):2083–9.
4. Fridell RA, Qiu D, Wang C, Valera L, Gao M. Resistance analysis of the hepatitis C virus NS5A inhibitor BMS-790052 in an in vitro replicon system. Antimicrob Agents Chemother. 2010;54(9):3641–50.
5. Wang C, Sun JH, O'Boyle 2nd DR, Nower P, Valera L, Roberts S, et al. Persistence of resistant variants in hepatitis C virus-infected patients treated with the NS5A replication complex inhibitor daclatasvir. Antimicrob Agents Chemother. 2013;57(5):2054–65.
6. Friborg J, Zhou N, Han Z, Yang X, Falk P, Mendez P, et al. In vitro assessment of re-treatment options for patients with hepatitis C virus genotype 1b infection resistant to daclatasvir plus asunaprevir. Infect Dis Ther. 2014.
7. Gane EJ, Stedman CA, Hyland RH, Ding X, Svarovskaia E, Symonds WT, et al. Nucleotide polymerase inhibitor sofosbuvir plus ribavirin for hepatitis C. N Engl J Med. 2013;368(1):34–44.

Emerging Antivirals in the Future

8

C. Nelson Hayes, Michio Imamura, and Kazuaki Chayama

Abstract

The safety and effectiveness of antiviral therapy for chronic hepatitis C has improved markedly with the introduction of direct-acting antiviral drugs and a concomitant decrease in interferon use. Although DAAs are potent antivirals, the emergence of resistance against DAAs has spurred the development of new drugs. Second-generation NS5A inhibitors have a higher genetic barrier compared to first-generation NS5A inhibitors and are highly effective against strains that are resistant to first-generation NS5A inhibitors. While new drug development has primarily focused on DAAs, another way to counter DAA resistance is to develop combination therapies that target host factors in addition to viral factors because it is more difficult for the virus to overcome changes in the host environment. For example, miravirsen targets host microRNA-122, which is highly expressed in hepatocytes and essential for viral replication. Emergence of resistance mutations in such therapies is very low. Therefore, combined use of DAAs with other drugs is expected in the future to achieve high SVR rates while minimizing the risk of resistance.

C.N. Hayes • M. Imamura
Department of Gastroenterology and Metabolism, Applied Life Sciences, Institute of Biomedical and Health Sciences, Hiroshima University, 1-2-3 Kasumi, Minami-ku, Hiroshima 734-8551, Japan

Liver Research Project Center, Hiroshima University, Hiroshima 734-8551, Japan

K. Chayama, MD, PhD (✉)
Department of Gastroenterology and Metabolism, Applied Life Sciences, Institute of Biomedical and Health Sciences, Hiroshima University, 1-2-3 Kasumi, Minami-ku, Hiroshima 734-8551, Japan

Liver Research Project Center, Hiroshima University, Hiroshima 734-8551, Japan

Laboratory for Digestive Diseases, Center for Genomic Medicine, RIKEN, Hiroshima 734-8551, Japan
e-mail: chayama@hiroshima-u.ac.jp

K. Chayama (ed.), *Hepatitis C Virus Treatment*,
DOI 10.1007/978-981-10-2416-0_8

Keywords
Chronic hepatitis C • Second-generation NS5A inhibitors • Non-nucleoside polymerase inhibitors • Miravirsen • Resistance-associated variants

8.1 Abbreviations

DAA	Direct-acting antiviral
HCV	Hepatitis C virus
NI	Nucleoside inhibitor
NNI	Non-nucleoside inhibitor
PEG-IFN	Pegylated interferon
PI	Protease inhibitor
SVR	Sustained viral response

8.2 Introduction

Direct-acting antiviral (DAA) drugs strongly inhibit replication of hepatitis C virus (HCV) by directly targeting essential viral proteins. Triple combination therapy with pegylated interferon (PEG-IFN) plus ribavirin (RBV) and a protease inhibitor, e.g., telaprevir [1], simeprevir [2], or vaniprevir [3], is currently used in Japan for treatment of genotype 1 chronic HCV infection. However, several interferon-free all-oral DAA combination therapies have recently been approved, including daclatasvir plus asunaprevir therapy [4] and sofosbuvir plus ledipasvir [5] for treatment of genotype 1 infections and sofosbuvir plus RBV for treatment of genotype 2 infections [6]. DAAs have potent antiviral effects, and each of these therapies has been shown to have high safety and efficacy. However, drug resistance has nonetheless occurred in some patients, leading to treatment failure and raising questions about the best approach to re-treating these patients. Fortunately, a variety of new drugs are currently under development. Given the large number of DAAs in various stages of clinical development, the number of treatment options is only expected to increase, and effective treatment for all patients is a major goal. This chapter outlines current drug resistance challenges and discusses trends in ongoing and future drug development.

8.2.1 NS3/NS4A Protease Inhibitors

The first DAAs to be approved were the protease inhibitors telaprevir and boceprevir. The 9.6 kb HCV RNA genome contains a single open reading frame coding for a single 3000-amino-acid polypeptide, which must then be cleaved into three structural and six nonstructural proteins. Cellular proteases cleave the three structural proteins, which include the core protein and two envelope proteins, and the non-structural proteins are cleaved at four sites by the virally encoded proteases NS2 and NS3 along with the NS4A cofactor. Telaprevir mimics the carboxy-terminal region of the HCV NS3/NS4 serine protease [7] and interferes with viral replication by preventing cleavage of the polyprotein. While telaprevir triple therapy improves SVR rates to around 70 %, aside from its inconvenient thrice-daily dosing regimen, the therapy is associated with a high frequency of adverse events, including pruritus, rash, and nausea [8], and has been reported to lead to treatment discontinuation in 18 % of patients [9]. Clinical trials in Japan reported comparable SVR rates but a higher incidence of adverse events, due in part to the relatively high fixed dose relative to body weight in Japanese patients [10].

The defining characteristic of DAAs is their high target specificity for viral proteins, but a consequence of this specificity is the relatively low barrier to resistance it presents to this highly adaptable RNA virus. Inter-genotypic variation in the NS3 domain restricts the use of first-generation protease inhibitors to HCV genotype 1 [11], but even within genotype 1, resistance occurs more frequently in genotype 1a than 1b due to a synonymous codon at R155 that reduces the number of nucleotide changes needed to achieve a favorable amino acid substitution [12]. Compensatory mutations such as V36M that restore viral fitness may also occur, allowing the virus to compete with wild-type virus in the absence of the drug [13], and cross-resistance prevents the use of related protease inhibitors.

8.2.2 Second-Wave Protease Inhibitors

Second-wave PIs, such as simeprevir, asunaprevir, faldaprevir, and paritaprevir, attempt to overcome these problems by increasing the barrier to resistance and broadening the antiviral activity to other genotypes. Although the historically difficult-to-treat genotype 1 is the most prevalent genotype worldwide with 46.2 % of cases, it has also received the most focus in drug development efforts, whereas genotype 4, which is overrepresented in low-income countries, remains difficult to treat with current therapies [14]. Another goal of second-wave PIs is to improve patient compliance by reducing the dosing schedule and reducing side effects [15]. In light of these improvements, telaprevir and boceprevir should be avoided as a first-line treatment.

Simeprevir has been approved in the USA (150 mg dose) and Japan (100 mg dose) for use in triple therapy with PEG-IFN and ribavirin. Simeprevir is a once-daily macrocyclic PI active against genotypes 1, 2, 4, 5, and 6 [16]. Simeprevir triple therapy achieves SVR rates of up to approximately 80 % [17–19], with similar incidence and severity of adverse events to PEG-IFN and ribavirin alone. With response-guided therapy, up to 96 % of prior relapsers were reported to achieve SVR, but viral breakthrough or relapse was common [20, 21]. Despite these promising results, the simeprevir-resistant Q80K mutation occurs frequently (9–48 %) in genotype 1a patients [22], potentially requiring screening of genotype 1a patients for the Q80 mutation and an alternative therapy if necessary.

8.2.3 Second-Generation Protease Inhibitors

Second-generation PIs attempt to go a step further and provide pan-genotypic activity against all HCV genotypes as well as resistance mutations affecting first-generation PIs. Grazoprevir (MK-5172) is a much anticipated once-daily second-generation PI currently undergoing advanced clinical testing. Grazoprevir is not sensitive to most variants affecting first-generation inhibitors and has a higher barrier to resistance, with SVR rates ranging from 89 % to 100 % [23]. In the randomized, open-label phase II C-WORTHY clinical trial, treatment-naive HCV genotype 1 patients were treated with grazoprevir plus the NS5A inhibitor elbasvir (MK-8742) with or without ribavirin for 12 weeks [24]. The SVR12 rate was 93 % for patients without ribavirin and 93 % with ribavirin. While no patients discontinued due to adverse events, virological failure occurred in 4 % of patients due to emergence of resistance-associated variants against grazoprevir or elbasvir. In another C-WORTHY study examining treatment duration in treatment-naive cirrhotic patients and prior null responders, SVR12 rates up to 100 % were achieved among prior null responders treated for 18 weeks with grazoprevir, elbasvir, and ribavirin [25]. The regimen is now undergoing phase III clinical testing in the C-Edge series of studies examining safety and efficacy of the therapy in a variety of difficult-to-treat patient populations, including those with HIV coinfection, chronic kidney disease, severe liver damage, or blood disorders. Results of a phase III C-Edge HIV coinfection study involving 12 weeks of grazoprevir plus elbasvir therapy indicate that 96 % of patients achieved SVR12, including all 35 patients with cirrhosis [26], while several patients relapsed, in two cases due to reinfection. Adverse events were mild and included fatigue, headache, and nausea. Improvements in drug development will continue to extend the reach of DAA therapy to patients with previously unmet treatment needs, although achieving 100 % SVR may prove challenging.

8.3 Second-Generation NS5A Inhibitors

NS5A inhibitors are among the most effective antivirals available, with picomolar efficacy and pan-genotypic activity, and might serve as the backbone of future HCV therapies. However, first-generation NS5A inhibitors are highly vulnerable to resistance variants, such as L31M and Y93H. Compared to the first-generation NS5A inhibitors daclatasvir and ledipasvir, second-generation inhibitors under development, such as ACH-3120, GS-5816, and MK-8742, have a higher genetic barrier and are expected to be highly effective against resistance mutations that affect first-generation drugs. Tables 8.1 and 8.2 show sensitivity of first- and second-generation NS5A inhibitors to wild-type and first-generation resistance mutations for genotype 1a and 1b using an HCV replicon [27]. The second-generation NS5A inhibitor ACH-3120 has been shown to have overall higher efficacy against NS5A mutations compared to ledipasvir and daclatasvir. For example, ledipasvir and daclatasvir show EC50 fold changes of 119 and 18 against NS5A-L31V variants relative to

Table 8.1 NS5A inhibitor susceptibility to amino acid substitutions in genotype 1a (modified from reference 7 [27])

Genotype 1a	ACH-3120		Ledipasvir		Daclatasvir	
NS5A amino acid substitution	EC50 (nM)	EC50 Fold change	EC50 (nM)	EC50 Fold change	EC50 (nM)	EC50 Fold change
Wild type	0.012	1	0.0061	1	0.015	1
M28T	0.27	23	0.0093	15	2.8	187
M28V	0.013	1	0.016	3	0.0058	0.39
Q30E	0.85	71	20	3279	61	4067
Q30H	0.011	1	0.63	103	2.5	167
Q30K	0.75	63	66	10,819	83	5533
Q30R	0.041	3	2.8	459	3.0	200
L31M	0.019	2	17	2787	1.9	127
L31V	0.016	1	2.4	393	11	733
P32L	0.095	2	0.97	41	1.5	29
H58D	0.1	8	–	–	1.9	127
Y93H	61	5083	30	4918	17	1133
K24R-Q30R	2.4	200	23	3770	17	1133

Table 8.2 NS5A inhibitor susceptibility to amino acid substitutions in genotype 1b (modified from reference 7 [27])

Genotype 1b	ACH-3120		Ledipasvir		Daclatasvir	
NS5A amino acid substitution	EC50 (nM)	EC50 Fold change	EC50 (nM)	EC50 Fold change	EC50 (nM)	EC50 Fold change
Wild type	0.0040	1	0.00077	1	0.0030	1
L28M	0.0045	1	0.0034	4	0.0031	1
L31F	0.0090	1	0.0077	3	0.046	5
L31M	0.0030	1	0.011	14	0.0094	3
L31V	0.0026	1	0.092	119	0.055	18
P32L	0.0025	1	0.0074	10	0.019	6
Y93H	0.0061	2	2.1	2727	0.12	40
Y93N	0.016	4	2.7	3506	0.23	77
L28M-Y93H	2.4	600	199	258,441	7.8	2600
L31M-Y93H	0.047	12	210	272,727	163	54,333
L31V-Y93N	0.95	238	256	332,467	265	88,333
P58A-Y93H	0.047	12	22	28,571	5.9	1967
P58S-T64A-Y93H	0.39	98	2.8	3636	1.1	467

wild type, whereas ACH-3102 shows no change in sensitivity. Similarly, ledipasvir and daclatasvir show EC50 changes of 2727 and 40 against the NS5A-Y93H wild type, while ACH-3102 shows only a slight reduction in sensitivity with a fold change of 2. Even against the double variant NS5A-L31M-Y93H, ACH-3120 remains relatively sensitive, with an EC50 fold change of 12, whereas ledipasvir and daclatasvir become almost completely ineffective, with fold changes of 272,727 and 54,333, respectively. This result suggests that treatment with a second-generation NS5A inhibitor may be an effective re-treatment option for patients who experienced treatment failure with daclatasvir and asunaprevir therapy due to NS5A-L31 and NS5A-Y93 mutations. Clinical trials are currently underway to examine the combination of ACH-3120 and sofosbuvir. In a phase 2 clinical trial, 36 treatment-naive genotype 1 HCV patients were treated with 50 mg ACH-3120 and 40 mg sofosbuvir once daily for 6 or 8 weeks [28]. All patients achieved SVR regardless of treatment period. In the future, ACH-312 and sofosbuvir combination therapy should be examined as a potential re-treatment option for patients who failed to clear the virus during daclatasvir and asunaprevir therapy.

8.4 Non-nucleoside Polymerase Inhibitors

Although both targeting the HCV RNA-dependent RNA polymerase, the polymerase inhibitor DAAs are divided into two drug classes, nucleoside and non-nucleoside inhibitors, that target different steps in RNA synthesis and have different mechanisms and resistance profiles. Following the success of the nucleoside inhibitor sofosbuvir, clinical testing of non-nucleoside inhibitors such as beclabuvir has begun. Beclabuvir, which inhibits polymerase activity by targeting the thumb 1 domain of the NS5B polymerase, has exhibited pan-genotypic antiviral activity against genotypes 1, 3, 4, 5, and 6 in vitro [29]. In a phase II clinical trial, 66 patients were treated for 12 or 24 weeks with beclabuvir in combination with daclatasvir (NS5A inhibitor) and asunaprevir (protease inhibitor), resulting in an SVR rate of 92 % [30]. However, beclabuvir is susceptible to mutations at NS5B-A421 and NS5B-P495 [31, 32]. A breakthrough occurred in genotype 1a patients with treatment-emergent NS3-R155K + NS5A-Q30R-L31M + NS5B-P495L mutations, and NS3-V36M-R155K+ NS5A-M28A-Q30R-H58P + NS5B-P495L mutations emerged at the time of relapse in a patient with preexisting NS3-V36M + NS5A-H58P mutations. In a phase III clinical trial, an SVR rate of 98 % (81/83) was observed for treatment-naive genotype 1b patients, and an SVR rate of 100 % (28/28) was observed for previously treated patients after 12 weeks of treatment [33]. Even among genotype 1a patients, SVR rates of 90 % (206/229) and 85 % (64/75) were observed in treatment-naive patients and previously treated patients, respectively. Furthermore, even genotype 1b patients with preexisting NS5A-L31I/M or NS5A-Y93H mutations were able to achieve SVR (Table 8.3). In an HCV replicon study involving genotype 1b NS5A-L31M-Y93H double mutants, no inhibitory effect was observed for daclatasvir and asunaprevir alone, but the three-way combination of daclatasvir, asunaprevir, and beclabuvir effectively suppressed

Table 8.3 SVR rates for combination therapy with daclatasvir, asunaprevir, and beclabuvir in patients with NS5A amino acid mutations (Modified from [33])

Genotype	NS5A amino acid substitution	Treatment naive ($n=312$)	Previously treated ($n=103$)
Genotype 1a	M28L/I/T/V	12/17 (71 %)	8/9 (89 %)
	Q30H/R	0/5 (0 %)	1/1 (100 %)
	L31M	2/2 (100 %)	2/2 (100 %)
	Y93H/C	1/2 (50 %)	0
Genotype 1b	L28M/V	1/1 (100 %)	1/1 (100 %)
	R30Q	3/3 (100 %)	1/1 (100 %)
	L31I/M	3/3 (100 %)	1/1 (100 %)
	Y93H	6/6 (100 %)	3/3 (100 %)

HCV replication [34]. This result suggests that addition of beclabuvir might be effective in re-treating patients who fail to respond to daclatasvir and asunaprevir therapy. On the other hand, the frequency of resistance mutations affecting non-nucleoside polymerase inhibitors and their effects on combination therapy are not well understood and must be further examined to determine the safest course of treatment for these patients.

8.5 Paritaprevir/Ritonavir, Ombitasvir, and Dasabuvir

Although many sofosbuvir-based therapies are being evaluated, several clinical trials have examined AbbVie's alternative DAA combination therapy consisting of paritaprevir with ritonavir (ABT-450/r, a protease inhibitor), ombitasvir (ABT-267, an NS5A inhibitor), and dasabuvir (ABT-333, an NNI polymerase inhibitor), with or without ribavirin. In a phase III clinical trial, treatment-naive genotype 1 patients who were treated with paritaprevir/r, ombitasvir, dasabuvir, and ribavirin for 12 weeks achieved an SVR12 rate of 96 % (genotype 1a, 95 %; genotype 1b, 98 %) [35]. In another phase III trial, 99.5 % of genotype 1b patients and 97 % of genotype 1a patients achieved SVR12 (36).

The combination therapy also showed promise for patients with cirrhosis or non-response to prior interferon therapy. In a phase III trial, genotype 1 patients who failed to achieve SVR during prior interferon therapy were treated with paritaprevir/r, ombitasvir, dasabuvir, and ribavirin for 12 weeks and achieved an SVR rate of 96 % [37]. In a phase III clinical study of patients with Child-Pugh class A compensated cirrhosis who were treated with paritaprevir/r, ombitasvir, dasabuvir, and ribavirin for 12 or 24 weeks, 91 % of patients in the 12-week arm achieved SVR, and 96 % of patients in the 24-week arm achieved SVR [38].

In December 2014, the FDA approved the Viekira Pak (co-formulated ombitasvir, paritaprevir, and ritonavir co-packaged with dasabuvir tablets) for treatment of genotype 1 infection, including patients with compensated cirrhosis. In September 2015, Japan approved AbbVie's VIEKIRAX (paritaprevir/ritonavir co-formulated with ombitasvir) for treatment of genotype 1 HCV infection, partly on the basis of results of the GIFT-1 clinical phase III clinical trial in which 94 % of non-cirrhotic patients and 91 % of cirrhotic patients were able to achieve SVR12 [39]. These approvals, along with the FDA's approval of Gilead's Harvoni (ledipasvir co-formulated with sofosbuvir) for genotype 1 in October 2014 and its expanded approval for genotypes 4, 5, and 6 in November 2015, signal an important trend toward fixed, once-daily dosing. While this change should simplify therapy implementation and improve patient compliance, it also limits the discretion of physicians to adjust the dosage or substitute an alternative DAA in response to patient needs, especially among cirrhotic patients. Shortly after approving the Viekira Pak, the FDA warned of serious liver injury or death in some patients with cirrhosis. While therapy was contraindicated or not recommended for many of these patients, treating cirrhotic patients remains a priority, and early discontinuation of DAA therapy could promote antiviral resistance. Interferon-free therapy for patients with advanced liver disease is a recent and unprecedented development, and the long-term outcomes and risks have yet to be determined.

8.6 Host Factor-Targeting Antivirals

8.6.1 Entry Inhibitors

While DAAs have improved treatment prospects for most patients, their safety and effectiveness is less clear in immunocompromised or coinfected patients, as well as patients with advanced liver disease. Prevention of graft reinfection in patients who receive liver transplantation is another concern that may require alternative or complementary approaches. These patients may benefit from HCV entry inhibitors, which interfere with early interactions between the HCV envelope proteins and host factors, by disrupting attachment to hepatocyte receptors or by interfering with post-binding events or viral fusion [40]. Saikosaponins, particularly SSb2, derived from *Bupleurum kaoi* root, have been shown to prevent HCV entry by targeting HCV E2 and inhibiting viral attachment and fusion events [41]. Synergistic interactions between DAAs and host-targeting agents such as antibodies against CD81, SR-B1, or CLDN1 might also help to improve efficacy in difficult-to-treat patients while reducing toxicity [42].

8.6.2 Cyclophilin A Inhibitors

While entry inhibitors prevent entry of HCV into uninfected cells, cyclophilin inhibitors act against already infected cells by disrupting the interaction between HCV NS5A and the HCV replication complex [43, 44]. Addition of the cyclophilin inhibitor SCY-635 was shown to restore interferon-stimulated gene activity in HCV-infected cells by reducing phosphorylation of two negative regulators of ISG activity, PKR and eIF2α [45]. Although the mechanism of action of cyclophilin inhibitors is incompletely understood, Chatterji et al. showed that cyclophilin A and NS5A are essential for the creation of double-membrane vesicles required for HCV RNA replication [46].

8.6.3 Silymarin

The milk thistle extract silymarin is an antioxidant with hepatoprotective effects on the liver by promoting hepatocyte regeneration and reducing inflammation and fibrogenesis [47] and has long been used for treatment of *Amanita phalloides*-induced liver failure [48]. High-dose intravenous silymarin has been successfully used to treat HCV patients who fail to respond to PEG-IFN plus ribavirin therapy [49–51].

8.6.4 Miravirsen

Highly expressed in the liver, the microRNA miR-122 plays an essential role in HCV replication and presents a potential host antiviral drug target [52]. Miravirsen, a 15-nucleotide locked nucleic acid (LNA), is a modified phosphorothioate antisense oligonucleotide that binds to miR-122 and inhibits its function [53]. Miravirsen has shown pan-genotypic inhibitory effects on replication against genotypes 1 through 6. MiR-122 binds both to the 5′UTR (S1 and S2) [54] and to the 3′UTR (S3) [55] of the HCV RNA genome. Long-term HCV replicon studies of miravirsen activity have not revealed mutations in the miR-122-binding region [56]. Efficacy of miravirsen against chronic hepatitis C patients has been examined in a phase II clinical trial. In this study, 36 genotype 1 patients were assigned to receive placebo or 3, 5, or 7 mg/kg of miravirsen administered weekly by subcutaneous injection for 5 weeks, and HCV RNA levels were monitored for 18 weeks [57]. In the miravirsen-treated patients, blood HCV RNA titers were significantly reduced in a dose-dependent manner that persisted even after the end of therapy. The average HCV RNA reduction was 0.4 log IU/mL in the placebo group, 1.2 log IU/mL ($P=0.01$) in the 3 mg/kg group, 2.9 log IU/mL ($P=0.003$) in the 5 mg/kg group, and 3.0 log IU/mL ($P=0.002$) in the 7 mg/kg group. In addition, after 14 weeks of follow-up, HCV RNA became undetectable in four patients in the 7 mg/kg group and one patient in the 5 mg/kg group. Following the end of treatment, HCV RNA increased again in some patients, but no mutations in the S1, S2, and S3 HCV miR-122-binding sites were observed. However, mutations at A4C and C3U in the full-length 5′UTR were recognized in these patients [56]. In an HCV replicon study, miravirsen sensitivity did not differ between A4C mutant and wild-type strains, but resistance in C3U mutants was sevenfold higher than wild type, which may be problematic during treatment with miravirsen. However, in combination with DAAs, IFN, or RBV, C3U mutations show comparable sensitivity to wild type (Table 8.4). The combination of miravirsen with these agents may help to effectively suppress emergence of resistance mutations.

Table 8.4 Susceptibility of various drugs to 5′UTR substitutions (Modified from [56])

Drug	Fold change relative to wild type	
	A4C substitution	C3U substitution
Miravirsen	1	7
Telaprevir	2	1
VX-222	1	1
Daclatasvir	ND	1
IFN-α2b	ND	1
Daclatasvir	2	0.3
Ribavirin	ND	1

ND no data

8.7 Conclusions

Development of new HCV drugs unaffected by resistance-associated variants that limit current DAA drugs is anticipated. In principle, an SVR rate for treatment of chronic hepatitis C approaching 100 % is possible. However, a number of issues remain with respect to the emergence of resistance mutations, re-treatment following treatment failure, and treatment of patients with renal function decline or decompensated cirrhosis. Although many new drugs are under development, recent clinical trials have focused on shortening the duration of therapy, identifying effective DAA combinations for genotype 1 as well as other genotypes, improving resistance and safety profiles, and evaluating the need for ribavirin. The recent trend toward co-formulated once-daily fixed-dose tablets simplifies the treatment landscape and should improve patient compliance but at the cost of reduced flexibility and potentially greater risk for cirrhotic patients. With the development of new drugs and more effective treatments, it is hoped that all patients with chronic hepatitis C can be successfully treated.

References

1. Kumada H, Toyota J, Okanoue T, Chayama K, Tsubouchi H, Hayashi N. Telaprevir with peginterferon and ribavirin for treatment-naive patients chronically infected with HCV of genotype 1 in Japan. J Hepatol. 2012;56(1):78–84.
2. Hayashi N, Izumi N, Kumada H, Okanoue T, Tsubouchi H, Yatsuhashi H, et al. Simeprevir with peginterferon/ribavirin for treatment-naive hepatitis C genotype 1 patients in Japan: CONCERTO-1, a phase III trial. J Hepatol. 2014;61(2):219–27.
3. Hayashi N, Mobashery N, Izumi N. Vaniprevir plus peginterferon alfa-2a and ribavirin in treatment-experienced Japanese patients with hepatitis C virus genotype 1 infection: a randomized phase II study. J Gastroenterol. 2015;50(2):238–48.
4. Kumada H, Suzuki Y, Ikeda K, Toyota J, Karino Y, Chayama K, et al. Daclatasvir plus asunaprevir for chronic HCV genotype 1b infection. Hepatology. 2014;59(6):2083–9.
5. Mizokami M, Yokosuka O, Takehara T, Sakamoto N, Korenaga M, Mochizuki H, et al. Ledipasvir and sofosbuvir fixed-dose combination with and without ribavirin for 12 weeks in

treatment-naive and previously treated Japanese patients with genotype 1 hepatitis C: an open-label, randomised, phase 3 trial. Lancet Infect Dis. 2015;15(6):645–53.
6. Omata M, Nishiguchi S, Ueno Y, Mochizuki H, Izumi N, Ikeda F, et al. Sofosbuvir plus ribavirin in Japanese patients with chronic genotype 2 HCV infection: an open-label, phase 3 trial. J Viral Hepat. 2014;21(11):762–8.
7. Perni RB, Almquist SJ, Byrn RA, Chandorkar G, Chaturvedi PR, Courtney LF, et al. Preclinical profile of VX-950, a potent, selective, and orally bioavailable inhibitor of hepatitis C virus NS3-4A serine protease. Antimicrob Agents Chemother. 2006;50(3):899–909.
8. Jacobson IM, McHutchison JG, Dusheiko G, Di Bisceglie AM, Reddy KR, Bzowej NH, et al. Telaprevir for previously untreated chronic hepatitis C virus infection. N Engl J Med. 2011;364(25):2405–16.
9. Sherman KE, Flamm SL, Afdhal NH, Nelson DR, Sulkowski MS, Everson GT, et al. Response-guided telaprevir combination treatment for hepatitis C virus infection. N Engl J Med. 2011;365(11):1014–24.
10. Chayama K, Hayes CN, Yoshioka K, Moriwaki H, Okanoue T, Sakisaka S, et al. Accumulation of refractory factors for pegylated interferon plus ribavirin therapy in older female patients with chronic hepatitis C. Hepatol Res: Off J Jpn Soc Hepatol. 2010;40(12):1155–67.
11. Halfon P, Locarnini S. Hepatitis C virus resistance to protease inhibitors. J Hepatol. 2011;55(1):192–206.
12. Lok AS, Gardiner DF, Lawitz E, Martorell C, Everson GT, Ghalib R, et al. Preliminary study of two antiviral agents for hepatitis C genotype 1. N Engl J Med. 2012;366(3):216–24.
13. Wyles DL. Beyond telaprevir and boceprevir: resistance and new agents for hepatitis C virus infection. Top Antivir Med. 2012;20(4):139–45.
14. Messina JP, Humphreys I, Flaxman A, Brown A, Cooke GS, Pybus OG, et al. Global distribution and prevalence of hepatitis C virus genotypes. Hepatology. 2015;61(1):77–87.
15. Wendt A, Adhoute X, Castellani P, Oules V, Ansaldi C, Benali S, et al. Chronic hepatitis C: future treatment. Clin Pharmacol: Adv Appl. 2014;6:1–17.
16. Rosenquist A, Samuelsson B, Johansson PO, Cummings MD, Lenz O, Raboisson P, et al. Discovery and development of simeprevir (TMC435), a HCV NS3/4A protease inhibitor. J Med Chem. 2014;57(5):1673–93.
17. Jacobson IM, Dore GJ, Foster GR, Fried MW, Radu M, Rafalsky VV, et al. Simeprevir with pegylated interferon alfa 2a plus ribavirin in treatment-naive patients with chronic hepatitis C virus genotype 1 infection (QUEST-1): a phase 3, randomised, double-blind, placebo-controlled trial. Lancet. 2014;384(9941):403–13.
18. Manns M, Marcellin P, Poordad F, de Araujo ES, Buti M, Horsmans Y, et al. Simeprevir with pegylated interferon alfa 2a or 2b plus ribavirin in treatment-naive patients with chronic hepatitis C virus genotype 1 infection (QUEST-2): a randomised, double-blind, placebo-controlled phase 3 trial. Lancet. 2014;384(9941):414–26.
19. Forns X, Lawitz E, Zeuzem S, Gane E, Bronowicki JP, Andreone P, et al. Simeprevir with peginterferon and ribavirin leads to high rates of SVR in patients with HCV genotype 1 who relapsed after previous therapy: a phase 3 trial. Gastroenterology. 2014;146:1669.
20. Izumi N, Hayashi N, Kumada H, Okanoue T, Tsubouchi H, Yatsuhashi H, et al. Once-daily simeprevir with peginterferon and ribavirin for treatment-experienced HCV genotype 1-infected patients in Japan: the CONCERTO-2 and CONCERTO-3 studies. J Gastroenterol. 2014;49(5):941–53.
21. Zeuzem S, Berg T, Gane E, Ferenci P, Foster GR, Fried MW, et al. Simeprevir increases rate of sustained virologic response among treatment-experienced patients with HCV genotype-1 infection: a phase IIb trial. Gastroenterology. 2014;146(2):430–41.e6.
22. Schneider MD, Sarrazin C. Antiviral therapy of hepatitis C in 2014: do we need resistance testing? Antiviral Res. 2014;105c:64–71.
23. Gentile I, Buonomo AR, Borgia F, Zappulo E, Castaldo G, Borgia G. MK-5172: a second-generation protease inhibitor for the treatment of hepatitis C virus infection. Expert Opin Investig Drugs. 2014;23(5):719–28.

24. Sulkowski M, Hezode C, Gerstoft J, Vierling JM, Mallolas J, Pol S, et al. Efficacy and safety of 8 weeks versus 12 weeks of treatment with grazoprevir (MK-5172) and elbasvir (MK-8742) with or without ribavirin in patients with hepatitis C virus genotype 1 mono-infection and HIV/hepatitis C virus co-infection (C-WORTHY): a randomised, open-label phase 2 trial. Lancet. 2015;385(9973):1087–97.
25. Lawitz E, Gane E, Pearlman B, Tam E, Ghesquiere W, Guyader D, et al. Efficacy and safety of 12 weeks versus 18 weeks of treatment with grazoprevir (MK-5172) and elbasvir (MK-8742) with or without ribavirin for hepatitis C virus genotype 1 infection in previously untreated patients with cirrhosis and patients with previous null response with or without cirrhosis (C-WORTHY): a randomised, open-label phase 2 trial. Lancet. 2015;385(9973):1075–86.
26. Sulkowski MS, Sherman KE, Dieterich DT, Bsharat M, Mahnke L, Rockstroh JK, et al. Combination therapy with telaprevir for chronic hepatitis C virus genotype 1 infection in patients with HIV: a randomized trial. Ann Intern Med. 2013;159(2):86–96.
27. Patel D, Zhao Y, Fabrycki J, Yang G, Podos S, Kocinsk H, et al. P0805: achievement of SVR12 despite the presence of HCV variants resistant to first generation NS5A inhibitors in genotype-1 hepatitis C patients after 8-week therapy of ACH-3102 in combination with sofosbuvir. J Hepatol. 62:S636.
28. Gane E, Schwabe C, Mader M, Suri V, Donohue M, Huang M, et al. LP06: sustained virologic response after ACH-3102 and sofosbuvir treatment for 8 or 6 weeks: a phase 2 "proxy" study. J Hepatol. 62:S266.
29. Liu M, Tuttle M, Gao M, Lemm JA. Potency and resistance analysis of hepatitis C virus NS5B polymerase inhibitor BMS-791325 on all major genotypes. Antimicrob Agents Chemother. 2014;58(12):7416–23.
30. Everson GT, Sims KD, Rodriguez-Torres M, Hezode C, Lawitz E, Bourliere M, et al. Efficacy of an interferon- and ribavirin-free regimen of daclatasvir, asunaprevir, and BMS-791325 in treatment-naive patients with HCV genotype 1 infection. Gastroenterology. 2014;146(2):420–9.
31. Shi N, Hiraga N, Imamura M, Hayes CN, Zhang Y, Kosaka K, et al. Combination therapies with NS5A, NS3 and NS5B inhibitors on different genotypes of hepatitis C virus in human hepatocyte chimeric mice. Gut. 2013;62(7):1055–61.
32. Abe H, Hayes CN, Hiraga N, Imamura M, Tsuge M, Miki D, et al. A translational study of resistance emergence using sequential direct-acting antiviral agents for hepatitis C using ultra-deep sequencing. Am J Gastroenterol. 2013;108(9):1464–72.
33. Reddy KR, Bourliere M, Sulkowski M, Omata M, Zeuzem S, Feld JJ, et al. Ledipasvir and sofosbuvir in patients with genotype 1 hepatitis C virus infection and compensated cirrhosis: an integrated safety and efficacy analysis. Hepatology. 2015;62:79.
34. Friborg J, Zhou N, Han Z, Yang X, Falk P, Mendez P, et al. In vitro assessment of re-treatment options for patients with hepatitis C virus genotype 1b infection resistant to daclatasvir plus asunaprevir. Inf Dis Ther. 2014.
35. Feld JJ, Kowdley KV, Coakley E, Sigal S, Nelson DR, Crawford D, et al. Treatment of HCV with ABT-450/r-ombitasvir and dasabuvir with ribavirin. N Engl J Med. 2014;370(17):1594–603.
36. Ferenci P, Bernstein D, Lalezari J, Cohen D, Luo Y, Cooper C, et al. ABT-450/r-ombitasvir and dasabuvir with or without ribavirin for HCV. N Engl J Med. 2014;370(21):1983–92.
37. Zeuzem S, Jacobson IM, Baykal T, Marinho RT, Poordad F, Bourliere M, et al. Retreatment of HCV with ABT-450/r-ombitasvir and dasabuvir with ribavirin. N Engl J Med. 2014;370(17):1604–14.
38. Poordad F, Hezode C, Trinh R, Kowdley KV, Zeuzem S, Agarwal K, et al. ABT-450/r--ombitasvir and dasabuvir with ribavirin for hepatitis C with cirrhosis. N Engl J Med. 2014;370(21):1973–82.
39. Kumada H, Chayama K, Rodrigues Jr L, Suzuki F, Ikeda K, Toyoda H, et al. Randomized phase 3 trial of ombitasvir/paritaprevir/ritonavir for hepatitis C virus genotype 1b-infected Japanese patients with or without cirrhosis. Hepatology. 2015;62(4):1037–46.

40. Fofana I, Jilg N, Chung RT, Baumert TF. Entry inhibitors and future treatment of hepatitis C. Antiviral Res. 2014;104:136–42.
41. Lin LT, Chung CY, Hsu WC, Chang SP, Hung TC, Shields J, et al. Saikosaponin b2 is a naturally occurring terpenoid that efficiently inhibits hepatitis C virus entry. J Hepatol. 2015;62(3):541–8.
42. Xiao F, Fofana I, Thumann C, Mailly L, Alles R, Robinet E, et al. Synergy of entry inhibitors with direct-acting antivirals uncovers novel combinations for prevention and treatment of hepatitis C. Gut. 2015;64(3):483–94.
43. Hopkins S, Bobardt M, Chatterji U, Garcia-Rivera JA, Lim P, Gallay PA. The cyclophilin inhibitor SCY-635 disrupts hepatitis C virus NS5A-cyclophilin A complexes. Antimicrob Agents Chemother. 2012;56(7):3888–97.
44. Hopkins S, Gallay P. Cyclophilin inhibitors: an emerging class of therapeutics for the treatment of chronic hepatitis C infection. Viruses. 2012;4(11):2558–77.
45. Daito T, Watashi K, Sluder A, Ohashi H, Nakajima S, Borroto-Esoda K, et al. Cyclophilin inhibitors reduce phosphorylation of RNA-dependent protein kinase to restore expression of IFN-stimulated genes in HCV-infected cells. Gastroenterology. 2014;147(2):463–72.
46. Chatterji U, Bobardt M, Tai A, Wood M, Gallay PA. Cyclophilin and NS5A inhibitors, but not other anti-hepatitis C virus (HCV) agents, preclude HCV-mediated formation of double-membrane-vesicle viral factories. Antimicrob Agents Chemother. 2015;59(5):2496–507.
47. Feher J, Lengyel G. Silymarin in the prevention and treatment of liver diseases and primary liver cancer. Curr Pharm Biotechnol. 2012;13(1):210–7.
48. Mengs U, Pohl RT, Mitchell T. Legalon(R) SIL: the antidote of choice in patients with acute hepatotoxicity from amatoxin poisoning. Curr Pharm Biotechnol. 2012;13(10):1964–70.
49. Ferenci P, Scherzer TM, Kerschner H, Rutter K, Beinhardt S, Hofer H, et al. Silibinin is a potent antiviral agent in patients with chronic hepatitis C not responding to pegylated interferon/ribavirin therapy. Gastroenterology. 2008;135(5):1561–7.
50. Polyak SJ, Ferenci P, Pawlotsky JM. Hepatoprotective and antiviral functions of silymarin components in hepatitis C virus infection. Hepatology. 2013;57(3):1262–71.
51. Dahari H, Shteingart S, Gafanovich I, Cotler SJ, D'Amato M, Pohl RT, et al. Sustained virological response with intravenous silibinin: individualized IFN-free therapy via real-time modelling of HCV kinetics. Liver Int: Off J Int Assoc Study Liver. 2015;35(2):289–94.
52. Jopling CL, Yi M, Lancaster AM, Lemon SM, Sarnow P. Modulation of hepatitis C virus RNA abundance by a liver-specific MicroRNA. Science. 2005;309(5740):1577–81.
53. Li YP, Gottwein JM, Scheel TK, Jensen TB, Bukh J. MicroRNA-122 antagonism against hepatitis C virus genotypes 1–6 and reduced efficacy by host RNA insertion or mutations in the HCV 5′UTR. Proc Natl Acad Sci U S A. 2011;108(12):4991–6.
54. Jopling CL, Schutz S, Sarnow P. Position-dependent function for a tandem microRNA miR-122-binding site located in the hepatitis C virus RNA genome. Cell Host Microbe. 2008;4(1):77–85.
55. Henke JI, Goergen D, Zheng J, Song Y, Schuttler CG, Fehr C, et al. microRNA-122 stimulates translation of hepatitis C virus RNA. EMBO J. 2008;27(24):3300–10.
56. Ottosen S, Parsley TB, Yang L, Zeh K, van Doorn LJ, van der Veer E, et al. In vitro antiviral activity and preclinical and clinical resistance profile of miravirsen, a novel anti-hepatitis C virus therapeutic targeting the human factor miR-122. Antimicrob Agents Chemother. 2015;59(1):599–608.
57. Janssen HL, Reesink HW, Lawitz EJ, Zeuzem S, Rodriguez-Torres M, Patel K, et al. Treatment of HCV infection by targeting microRNA. N Engl J Med. 2013;368(18):1685–94.

MIX
Papier aus verantwortungsvollen Quellen
Paper from responsible sources
FSC® C105338

If you have any concerns about our products,
you can contact us on
ProductSafety@springernature.com

In case Publisher is established outside the EU,
the EU authorized representative is:
Springer Nature Customer Service Center GmbH
Europaplatz 3, 69115 Heidelberg, Germany

Printed by Libri Plureos GmbH
in Hamburg, Germany